BT COMMUNICATIONS TECHNOLOGY SERIES 2

Future mobile networks:
3G and beyond

Other volumes in this series:

Volume 1 **Carrier-scale IP networks: designing and operating Internet networks** P. Willis
Volume 2 **Future mobile networks: 3G and beyond** A. Clapton
Volume 3 **Voice over IP** R. Swale

Future mobile networks:
3G and beyond

Edited by

Alan Clapton

The Institution of Electrical Engineers

Published by: The Institution of Electrical Engineers, London,
United Kingdom

The Institution of Electrical Engineers,
Michael Faraday House,
Six Hills Way, Stevenage,
Herts. SG1 2AY, United Kingdom

British Library Cataloguing in Publication Data

Future mobile networks.
(BT communications technology series; no. 2)
1. Wireless communication systems 2. Mobile communication systems
I. Clapton, A.
621.3′ 82

ISBN 0 85296 983 X

Typeset in the UK by Mendez Ltd, Ipswich
Printed in the UK by T.J. International, Padstow

CONTENTS

PREFACE

Mobile systems in one guise or another have very much been in the news during the past year. This book is very much focused on the technology, research and ongoing developments in the mobile arena.

The following news items illustrate the scale involved within mobile developments today. This snapshot of a two-week period depicts how companies are forming partnerships to create and deploy these new technologies, for operators and others, to offer to the customer. It also clearly demonstrates the rapid convergence that is occurring between mobile and the Internet, which is a key element of the chapters within this edition.

> "The US Department of Justice has given clearance to the wireless joint venture of SBC Communications and Bell South Corporation — provided overlapping wireless properties are resolved. The companies plan to merge their domestic wireless operations to create the nation's second largest wireless company — a powerful single entity which will serve more than 18 million customers with 175 million potential customers in 34 states." 1/9/00

> "Swedish telecomms manufacturer Ericsson has announced that — along with BT alliance Japan Telecom (JT) — it has successfully completed the world's first field trial of IP-based voice (VoIP) over JT's chosen third generation platform, WCDMA (wideband code division multiple access)." 7/9/00

> "BT's Genie is creating the UK's first mass-market 'eCash' facility for Internet shopping and pre-pay mobile top-ups with the launch to Qdodge.com from October." 11/9/00

> "Ericsson and Microsoft have launched Ericsson Microsoft Mobile Venture AB, a joint venture to drive the mobile Internet by developing and marketing mobile e-mail solutions. The first solutions are expected to be on the market by the end of the year. The company is part of a broader strategic alliance between Ericsson and Microsoft." 12/9/00

> "Compaq Computer Corporation and Nokia have announced a global alliance to jointly develop and market end-to-end mobile Internet and Intranet solutions to enterprise customers." 12/9/00

> "Leicestershire Constabulary will be one of the first forces in the country to introduce a state-of-the-art digital radio service to enable emergency services around Britain to communicate with each other. The BT Airwave secure mobile radio communication service — believed to be the most advanced system of its kind in the world — will provide officers with a new national network to fight crime, using a range of communications tools." 12/9/00

The basic story line behind all these events is that the number of mobile customer numbers continues to grow substantially on a global scale. We are also starting to see the early convergence of mobile and the Internet, which, it is envisaged, will provide the next massive growth burst to the mobile market. We are also seeing 'time to market' being compressed, which introduces changes in the approach to research and in standards developments. The mobile customer base is also becoming more sophisticated and expects services to be available as they move around, just as they can access them in their home or office environment; these include the greater bandwidth and higher quality demanding services, such as video and large file transfers.

As well as the continued growth in the cellular mobile market-place, we have seen the emergence and introduction of new private mobile radio (PMR) systems tailored specifically for the demanding requirements of the emergency services and large fleet companies.

With this in mind, the contributors to this book, to whom I am indebted, have endeavoured to address the developments that are likely to occur in the near future as the drive continues towards those technology capabilities required to meet the expanding market expectation in this dynamic environment. The following paragraphs provide an introduction to the content of each chapter.

Research has formed the bedrock of mobility developments particularly in Europe through support of third generation initiatives by the European community (RACE, ACTS, etc). The time-scales for delivery are now increasingly being compressed and it is necessary to look at innovative ways in which research activities can be brought to focus rapidly and effectively. The first two chapters address two different approaches that are taking place to address this problem, specifically highlighting the developments that are taking place in collaborative activities within academic institutes. Chapter 1 addresses the work that has been undertaken by BT in conjunction with a number of leading technology universities, termed the Virtual University Research Initiative (VURI). This work has investigated a number of future developments in radio access and other areas. Chapter 2 addresses the development of the UK initiative to generate collaboration between a number of leading UK companies and the leading universities in the mobile field.

Innovative technology developments still form an essential feature in mobile systems developments. With the introduction of GSM it was impossible to foresee, for example, the use that customers would make of the GSM Short Message

Service. This is probably one of the most difficult areas to assess how technology developments are turned into products that can be sold to the customer. In reality, the mobile market has been driven by an ever-increasing array of products and capabilities and the successful operator is the one who can best turn those into the products that users really want. Chapter 3 identifies some of the product themes that we might expect may start being made available to customers with the introduction of 3G. In particular, the chapter highlights some of the unique mobile product opportunities such as those offered by geographic location capabilities.

In the mobile industry, standards have played a major part in creating a competitive environment that has enabled suppliers to provide equipment to operators in a true multi-vendor environment. This is particularly true in the case of the terminals where we have a large choice from an array of suppliers — consumers can now procure with the confidence that they will work on any compatible network in the world. Chapter 4 describes the standards creation forums and their roles in producing the 3G solutions of tomorrow.

Mobile technology developments have always used as a basis the convergence of mobility with some fixed network counterpart. GSM set this trend through basing its development on narrowband ISDN. The general packet radio service continued this trend through integration with Internet Protocols. 3G developments continued with higher speed access to the switched and packet-based GSM/GPRS networks. The next major drive in the mobile industry is to provide a higher degree of integration with the Internet enabling the true multimedia experience, including the opportunity for voice over IP (VoIP)- the so-called 'all IP' solution for 3G. Chapter 5 describes the architectural principles that are being addressed to provide the framework to support this solution and the developments to date in this important area.

Chapters 6 and 7 list the more detailed aspects of the solutions being put in place to deliver the specific developments that will be needed with the IP solution. A major aspect in this development is providing convergence with IP solutions while still maintaining the ability for users to roam across operators' networks based either on this solution, or on less mature 3G and 2G networks, in a seamless way. It is important in all these developments not to lose the major benefit that GSM has created on a global scale for the user, namely the ability to move around the world over compatible networks and simply turn on the mobile and communicate. Specifically the chapters address the service control and the call control developments that are necessary in the mobile piece. Chapter 6 identifies how the call/session control will make use of the advanced IP session initiation protocol (SIP) as its basic element for establishing the end-to-end multimedia services. Chapter 7 describes how the service control will build on the developments we have today and explore the new opportunities that IP and SIP will introduce in the future.

While the Internet environment brings to mobile unique opportunities, it also carries threats to its existing service delivery processes. GSM solutions have been built on standardised service capabilities delivered in a vanilla flavour by suppliers to operators. The Internet opens up opportunities for service development by an entrepreneurial community, creating applications quickly, which may succeed or fail. An important aspect of the evolving solutions for 3G is how to provide an application-hosting environment for third party providers, be they content or service provision. Chapters 8 and 9 address different compatible offerings as to how these solutions may be provided. Chapter 8 introduces the concepts of an open service architecture. This concept provides for application programming interfaces (APIs) to be created and deployed by network operators for use by third party providers. These APIs provide access to a number of the network-service-providing platforms in a simple and easily usable format. Chapter 9 addresses how mobile portals are being created to provide adaptive access mechanisms between mobile terminals and the Internet. The chapter addresses some of the mechanisms and capabilities that can be increasingly applied to make the user experience much more service rich and easy to use.

A major factor in the dynamic growth of the mobile market has been the diversity in supply of mobile terminals. This diversity allows a wide choice for customers taking up the service. Complexity has grown significantly over the years and a major feature in terminal development has been how the supplier can create a simple intuitive interface for the user. In effect, the mobile device is at the heart of any development. It is through the terminal that the user gets all their services delivered, so it is essential that the experience of that use is friendly, easy to use, and comfortable. Chapter 10 addresses some of the changes and developments that could be occurring in the near future to simplify and enhance the user experience.

When thinking of mobile developments the one unique high value feature is the radio interface. As we have seen in recent 3G licence auctions this is also one of the most costly parts as well. Any discussion of future mobile systems cannot be complete without a review of radio access and an overview of the changes and developments we can expect in the future. The very large sums of money being paid for spectrum make the radio resource a vital element of the overall system. It is therefore vital to make high-quality use of this resource while maximising the capacity available to sell to users. Chapter 11 addresses the on-going and future developments which could be expected for radio access.

In its initial stages of development, the convergence of mobile with the Internet is based on the further evolution of GPRS, as described in previous chapters. Already we are starting to see in research areas the next stages of development where further integration with the Internet will be managed through progressive stages. Chapter 12 identifies some of the mobility-enabled routing developments that are now being progressed within the Internet arena. These developments potentially will enable access between a mixture of mobile cellular and other environments, such as wireless LANs, etc. This starts to provide the ultimate vision of service availability in whatever environment users find themselves — be it wide area, corporate office or home.

The twin aims of increasing the mobile market and introducing ever more complex applications and developments make it all the more important to run trials on the various stages of these developments prior to full commercial launch. These trials allow suppliers, operators, and service providers to collaborate and build confidence in solutions enabling highly efficient services and applications to be deployed for customers at launch. Chapter 13 identifies some of the trial experiences that have been undertaken as an aid to developing 3G Internet solutions.

And finally, Chapter 14 takes an orthogonal view of another major mobility market. The majority of this book has concentrated on the convergence of cellular access mobility systems with the Internet. However, there is a very large user base for customised services that have very exacting requirements, namely police, fire and ambulance services. This chapter identifies how the TETRA professional mobile radio service can be evolved and converged with the Internet to offer greater choice to that user group while still maintaining their demanding service needs.

The production of this book has been an exciting project with which to be involved, not least because of the breathtaking speed at which the growth and breadth of mobile development is happening. I hope you find its contents interesting and thought provoking and perhaps, next time you use your mobile, you may stop and think what a marvel it is, and the opportunities that will be available in the near future.

Alan Clapton
Mobile Network Design, BT Wireless

CONTRIBUTORS

M C Bale, 3G Networks, BTexact Technologies

A R Beresford, Mobility Futures, BTexact Technologies

A Brookland, Mobile Solutions Development, BTexact Technologies

R Bloomfield, 3G Service Performance, BTexact Technologies

K M Brown, Mobility Solutions Development, Ignite Solutions

M D Cookson, Signalling Protocols, BTexact Technologies

R M Dennis, Mobility Futures, BTexact Technologies

C J Fenton, 3G Service Development and Trials, BT Wireless

J F Fisher, Mobile Strategy, BT Wireless

C D Friel, Service and Network Standards, BT Wireless

J Gil, Mobile Network Emulation, BT Wireless

K G Hall, Standards Policy, BT Wireless

J A Harmer, Terminals and Applications, BTexact Technologies

J W Harris, VoIP Network Solutions, BTexact Technologies

F Harrison, Technical Standards, BT Wireless

K A Holley, Terminal and Applications Standards, BT Wireless

W Johnston, Mobile Multimedia Design, BT Wireless

N C Lobley, UMTS Systems Design, BT Wireless

D W Lock, 3G Performance, BT Wireless

J G O Moss, Radio Design and Systems, BT Wireless

A W O'Neill, formerly Internet Protocol Suite, BTexact Technologies

D T Pratt, Mobile Solutions Development, BT Wireless

D Ralph, 3G Applications, BTexact Technologies

C G Shephard, Wireless Solutions, BTexact Technologies

D G Smith, Mobile Solutions Design, BT Wireless

R M Stretch, Signalling Protocols, BTexact Technologies

P R Tattersall, Airwave Future Services Strategy, BT Wireless

G Tsirtsis, formerly Internet Access, BTexact Technologies

1

VIRTUAL UNIVERSITY RESEARCH INITIATIVE ON MOBILITY

R M Dennis, A R Beresford and K M Brown

1.1 Introduction

The Virtual University Research Initiative (VURI) on Mobility is a corporately funded research project, which has looked at a wide range of topics and issues in the mobile radio and mobility area over several years. Although it has included other universities over this period, its success can be attributed to BT's long-term relationship with three UK universities — Birmingham [1], Bristol [2] and Oxford [3]. This relationship has provided a reservoir of skills and expertise, which have remained reasonably consistent throughout considerable changes in BT's research and development objectives in the mobility area.

The project has had a number of BT managers, and been associated with several different research and development units. This chapter will chart this relationship and its benefit to the business as well as looking at the future significance and direction of the VURI.

1.2 Background — Brief History of the VURI

In 1994, an audit of corporately funded research revealed a lack of expertise in the (then) new third-generation mobile technologies. The VURI was subsequently set up in 1995 to carry out research along lines suggested by both BT and the member institutions, and also act as a non-industrial expert group. The set-up of the VURI was also an experiment into a new way of organising university research — the universities and BT's laboratories at Adastral Park are all members of one 'virtual university' despite having a number of bases around the country. All VURI members have both individual projects and a commitment to working together in collaborative work.

The VURI project has been a useful test bed for assessing this style of managing large-scale, long-term research. The particular form of the VURI — made up of academics from different institutions together with BT researchers — has proved to be an effective way of generating reliable reports on a wide variety of technological issues. The range of expertise generates (often opposing) viewpoints on a variety of topics, either throwing up new areas of research or allowing identification of the underlying issues in a given area. The VURI has been successful in seeding research at BT's laboratories in the mobility area and is now starting to provide graduates who are both technically skilled and commercially aware.

Three universities were chosen to work together in initial phase of the VURI — Oxford, Birmingham and Bristol. Each institution has specific, yet complementary, strengths:

- Birmingham is very strong on the theoretical and analytical aspects of mobile communication, and signal propagation in particular;
- Oxford was motivated from a practical side, providing real-time measurements and analysis;
- Bristol combined elements of both, looking at the mobile system from a wider perspective, and using real data combined with theoretical predictions to build models of the performance of mobile networks in a variety of scenarios.

1.2.1 Original Vision — 'A Factor of One-hundred-Fold Increase in Capacity'

Initially, the vision for the VURI was to develop methods that would result in an increase in mobile network capacity between a factor of ten and one hundred in an economical and viable way. Some of the early projects focused on GSM networks, looking at issues such as cell planning, but, for the most part, attention was on different transmission techniques — in particular spread spectrum. It was clear that increasing the available bandwidth would immediately increase capacity, but the main interest was in finding ways to optimise the use of spectrum. In the light of the UMTS spectrum auctions and the vast prices which broadband spectrum is commanding, this is just as relevant today.

As the VURI research progressed, VURI members (including BT) were working together on projects which would be the key to the provisioning of third and fourth-generation mobile services, but which did not fit with the original vision. In 1999 a reappraisal of the VURI led to a new objective and, soon after, the responsibility for the project was handed over to the current custodians, the Mobility Futures team within BTexact Technologies at Adastral Park.

1.2.2 Current Vision — 'How Do We (BT) Best Provide Mobile Customers with Services that are Both Flexible and Reliable'

The current vision encompasses some of the original physical layer activities, but more importantly ensures the VURI will look at the bigger picture — advancing the technology used in mobile networks to provide services with increased flexibility and reliability.

1.3 Achievements to Date

The VURI's achievements fall into two categories — hard deliverables (including papers and reports), and influence (a sounding board that includes consultancy). Although the former is measurable, the latter is far more subjective and difficult to assess fully, given the degree of people movement in BT.

1.3.1 Influence

Apart from the numerous conference proceedings and panels, the VURI is acknowledged as playing a significant role in the UMTS standardisation process. The VURI was involved with the trials of manufacturers' equipment and provided input to BT's Third-Generation Licence bid team.

The Exploiting Mobility Conference, which was held at BT's laboratories at Adastral Park in 1998, was well received and succeeded in increasing the visibility of the VURI, providing a catalyst for several collaborative projects within the BT 'Exploiting Mobility' campaign.

1.3.2 Papers and Reports

The VURI to date has produced over 50 conference and journal papers; a list of papers together with abstracts is available from the BT VURI Web site [4]. Research has covered a wide range of topics, although the majority can be grouped into six broad categories:

- modelling and measurement of radio propagation in a variety of indoor and outdoor environments;
- adaptive antenna systems;
- mobile network systems modelling and measurement;
- theoretical analysis of electromagnetic diffraction and scattering;

- advanced modulation and coding schemes;
- UMTS enhancement techniques.

1.4 Current Focus

The nature of the research, and the extensive use of PhD students at its core, mean the VURI has a certain built-in inertia. This is not a barrier to its value, although care is needed to ensure the focus remains current and aligned with BT's vision and long-term business goals. The VURI's overall vision has to be concise and clearly communicable both inside and outside BT and the wider research community. The focus is constantly being reviewed and updated; the current focus is 'Ultimate mobility — freedom to roam with a single terminal'.

At present, the mobile industry is looking to software defined radio (SDR) [5] to solve some of the problems, given the wide variety of mobile networks across the world (see Appendix A for details). The work and expertise of the VURI will be used to assess and quantify the advantages and disadvantages of the mass-market SDR for both the consumer and the network operator. Fundamental questions and possible topics under consideration are likely to include the following.

- What effect does SDR have on quality of service and achievable data rates? Is SDR capable of some of the advanced transceiver architectures possible in custom hardware? What effect will this have on network capacity?
- When will SDR become cheaper/lighter/smaller than a hardware solution? How does additional functionality needed in the device (e.g. a PDA) alter the dynamics?
- What effect does SDR have on operators upgrading their networks? Will it increase or decrease the number of different types of network in operation?
- What percentage of customers would benefit sufficiently from SDR to invest in the new technology?
- What are the technology drivers that will underpin an impact in the various global markets?
- What factors can the BT Group and its partners use to differentiate its products and services?

With greater than one billion mobile subscribers predicted by 2003, BT and its alliance partners will be looking for differentiation based on service offering to gain its share of global mobile subscribers. The trade-off between an all-hardware device and a reconfigurable software solution needs to be understood — it is anticipated that the network requirements and service options for operators will directly reflect the approach taken. The new focus aims to forge stronger links between BT's research projects and the other VURI members, possibly combining this work with strategic suppliers or collaboration with one or more of the start-up companies active in the area. Ideally, this would form a 'knowledge staircase', with the VURI as a foundation with measurements, simulation and theory.

1.5 Relationship with the VCE

The Mobile VCE [6] is a consortium of twenty-three telecommunications operators and equipment manufacturers (including BTCellnet) which fund and co-ordinate research at seven UK universities. The goal is to investigate what a fourth-generation mobile network will look like in 2010 (see Chapter 2 for more detail on the VCE).

The VCE is important to the VURI, since it provides a wealth of knowledge and research that can be built on. However, it is important to manage VURI research carefully — to avoid duplication of work. The VURI research is co-ordinated with this in mind, building on the work of the Mobile VCE. The VURI has the advantage of being able to use the VCE work in a BT-specific context while avoiding the problems associated with the decision-making process inside a large and diverse industry body.

The VURI research programme will allow BT to differentiate its network and service offering in 2010 from other Mobile VCE consortium members. No conflict or duplication exists between the VURI and the VCE and it is BT's responsibility to ensure this position continues.

1.6 Business Benefit

The VURI is not unique in its collaboration with universities. Several corporately funded projects have large numbers of university contracts with more than one academic institution. BT has funded several 'Strategic University Research Initiatives' over the years, building a relationship with a single key academic institution in a particular research area. The VURI differs mainly because the relationship is a collaboration between four partners. BT looks to provide a vision, identity and strategy for the work, but also attempts to foster collaboration between the various universities independent of its own internal research. This allows the VURI to remain effective and continue to move the vision forward even during periods of uncertainty within BT resulting from, for example, re-organisations and short-term distractions. As with any long-term investment, the value of the VURI cannot be judged simply by looking at its perceived value at any one instance in time. The true business benefit of the VURI since 1995 should be considered as a combination of the following factors:

- quantifiable achievements and their impact on the issues of the day;
- cost of acquiring the expertise inside or outside of the business;
- time required to build and develop the equivalent in-depth competence internally;
- consistency of research focus independent of short-term business pressures;
- value of the relationship;
- value of recruitment potential from the VURI member universities;
- ease of access to academic and industry-wide contacts;
- credibility and perception of BT's competence in the area with internal and external customers.

One of the key benefits of a long-term relationship with the academic community is the opportunity to recruit high-quality, business-aware students. This has provided BT with invaluable expertise in the evaluation of prototype UMTS equipment and research in general.

The VURI also has a public relations role both internally and externally. Internally it provides a route for BT's researchers to work with world-class experts as their careers develop. Externally, visibility of BT's involvement with in-depth research adds credibility to its desired position as an industry leader.

1.7 2000/2001 Programme

Senior university researchers (see Appendix B) have provided the driving force over the years, although the VURI only directly funds a combination of Research Assistants (RAs) and PhD students at each university. The RAs take a more responsive and short-term role in addition to their involvement with the core research, which is done mainly by PhD students in three-year projects. This combination of consistency with the opportunity to address some of the more short-term problems allows BT to gain help with some of the 'fire-fighting' issues while minimising disruption to the underlying research work. Fine-tuning the focus on a regular basis provides longer-term control of the overall direction.

Recently, the programme has been extensively reviewed and plans put in place to update the focus over the next three years.

The primary expertise of the key individuals at the three universities is at the physical (radio channel) layer. Although this expertise is critical to the success of the VURI, the value of adding a new partner is actively being considered to augment the skill-set and maximise the value of the physical layer work. Improved awareness of the impact and requirements of services and applications could provide BT with a faster route from the fundamental research to real or perceived differentiation in the market-place. Several options are under consideration including non-UK institutions, possibly one in the USA with which BT already has a strategic relationship (e.g. MIT or Berkeley).

The case study (see Appendix C) is an example of the work continuing in the 2000/2001 annual programme. This work involves BT, Birmingham and Oxford, although Oxford is in the process of beginning a new programme of work that is still being defined. Bristol are currently looking at improvements achievable by organising the wireless infrastructure and the wired backbone network on a more global scale. In particular, they are researching self-organising network techniques, including the concept of situation awareness. The aim is to allow the wired and wireless network to make use of information both from base-stations and from supporting network nodes to enhance performance globally.

The majority of the work in the current programme fits in well with the suggested focus — however, a healthy debate remains as to whether the SDR focus is in fact too restrictive to encompass all of the VURI's potential contributions to the mobility area.

1.8 Future Direction

Given the approach already discussed in this chapter, much of the research work of the VURI in the medium-term (2000—2003) is already reasonably well defined. In the longer term the VURI will provide the foundation for fourth-generation systems and, perhaps more importantly, identify potentially disruptive technologies. SDR or re-configurability has already been discussed, but other areas include new and enhanced coding schemes, smart antennas and *ad hoc* mobile networks.

Wider visibility of the VURI and its work is still needed. For the 2000/2001 programme, the annual review meeting will enable effective academic peer review of work carried out. The recent addition of an external VURI Web site [4] (outside the BT firewall) will also increase the general academic access to VURI work.

BT is continuing to ensure the true potential of the VURI relationship is achieved, including realising the commercial potential of the some of the VURI's work. Projects like the channel sounder (Appendix C) are ripe for development by manufacturers under licence.

1.9 Conclusions

The VURI research methodology fits well with BT's current approach to technology acquisition, with less being done directly in-house and more resulting from collaborative work. This chapter has shown how the approach works in practice and demonstrates the importance of in-depth internal competence to maximise the advantage generated.

The VURI is a unique relationship that BT has had the foresight to build and maintain over several years. Realistically, funding of the VURI has to be corporate, and as such will always be vulnerable to short-term pressure. However, the VURI is not only justifiable because of its quality of research, but also its reservoir of specialists and expertise which can be used to complement BT's competence.

Appendix A

VURI Technology Update — Software Radio and Software Defined Radio

The software radio (SR) [5] revolution extends the migration from analogue to digital by liberating radio-based services from a dependency on hard-wired characteristics. In 1995, Mitola coined the term 'software radio' and effectively defined the whole area [7].

The essence of the SR concept is the ability to update the different functional components of a radio system dynamically and is achieved by configuring re-programmable hardware with software. The reconfiguration can be achieved by a software download over a radio link or via a wired network. Apart from the ability to offer multiband configurations, the approach can be used to upgrade the handset, e.g. when new standards are released. Taken further, SR can be used to adapt the hardware dynamically (e.g. change of codec) in an attempt to maintain quality-of-service targets or improve coverage. SR is not a new technology, but an evolution and convergence of digital radio and software technologies.

Providing reconfigurability after leaving the factory requires general-purpose hardware that can cope with the constraints of all the mobile radio communications standards. SR devices digitise signals directly from the antenna and all the processing is performed by software residing in high-speed digital signal processors (DSPs). Limitations in DSP technology prevent construction of SR terminals in the short term; however, significant reconfigurability can be gained using software defined radio (SDR) — performing digitisation at some stage after the reception and analogue processing. As technology progresses, an SDR handset will provide increasing functionality, ultimately culminating in an SR device.

Industry experts suggest SDR systems will be available as early as 2003 in both base-station and handset configurations. Although BT is unlikely to build handsets or base-stations, it is essential to be aware of the evolution of programmable hardware because of the impact it will have on increasing the flexibility and adaptability of the network. With global procurement based on a 'vanilla' set of programmable building blocks, BT needs to understand the capabilities and limitations of this technology in order to maintain its ability to differentiate, based on the relationship between the network and the service or application.

Appendix B

VURI Member Institutions — the Key University Researchers

The Communications Group [3] within the Department of Engineering Science (headed by Professor David Edwards), at Oxford University, is broadly split into three areas — Future Radio Systems, Advanced RF and Microwave Devices, and Radar and Antenna Systems. Combined, these three groups provide the VURI insight into propagation measurement and modelling, adaptive antennas, and application of advanced materials in communications systems.

Professor David Edwards returned to academia in 1985 after 12 years with BT. He has contributed to more than 150 publications in the fields of radio and optical communications systems, electromagnetics and signal processing. Currently he is the Professor of Engineering Science at the University of Oxford. He has been in receipt of a number of awards for his research work and currently his interests cover mobile radio communications, free-space optical communications, high-temperature superconducting components for communications, and imaging techniques for communications and medical applications. He is a Fellow of the IEE and the Royal Astronomical Society.

The Communications Engineering Group [1] (where Costas Constantinou heads the radio wave propagation research activity), within the School of Electronic and Electrical Engineering, Birmingham, provides the VURI with expertise on propagation measurement and fundamental electromagnetic propagation theory as well as dynamic system control and network issues.

Costas Constantinou was born in Famagusta, Cyprus in 1964. He received his BEng (Hons) in electronic and communications engineering and PhD in electronic and electrical engineering degrees from the University of Birmingham, in 1987 and 1991, respectively. He is a Fellow of the IEE and the Royal Astronomical Society.

In 1989 he joined the School of Electronic and Electrical Engineering at the University of Birmingham as a full-time lecturer and, subsequently, as a senior lecturer. He currently heads the radio wave propagation research activity in the communications engineering research group. His research interests include optics, electromagnetic theory, electromagnetic scattering and diffraction, electromagnetic measurement, radio wave propagation modelling, mobile radio, and future communications networks architectures.

The Centre for Communications Research [2] (where Mark Beach leads the CDMA and adaptive antennas research programmes), within the Electronic and Electrical Engineering Department at the University of Bristol, has teams working in RF engineering and wireless communications systems. Currently, Bristol provides the VURI with expertise on situation awareness in self-organising networks and mobile network system simulation and modelling.

Mark Beach received his PhD in 1989 from the University of Bristol for work on adaptive antennas for multiple spread spectrum signal sources, primarily targeted towards GPS receiver technology. Post-doctoral research at Bristol included work regarding the application of adaptive antenna techniques to mobile cellular networks for which the research team received the IEEE Neal Shepherd memorial prize in 1990. Since August 1990, he has been engaged as a member of lecturing staff at Bristol and leads the CDMA and adaptive antennas research programmes within the Centre for Communications Research. In particular, he has led Bristol's activities on smart antennas under European funding from RACE, ACTS and now IST. This includes projects such as TSUNAMI, AWACS and SATURN, as well as helping the CEC launch the First European Colloquium on reconfigurable radio systems and networks in March 1999, from where projects such as TRUST were conceived. At present he holds the post of Reader in Communication Systems at the University of Bristol. He is also a serving member of the IEEE 8 Professional Group on Radio Communication Systems.

Appendix C

VURI Case Study — Radio Channel Photography

If the BT Group and its alliances are to maximise return from their investment in third-generation (3G) mobile telephone licences, it must produce a range of services which derive benefit from the high data rates available. Voice calls can already be carried on the existing second generation (2G) networks. 3G can only attract the volume of subscribers required by offering new and improved services and applications to provide the differentiation. To achieve the capacity and coverage required for high data rate services on 3G networks, careful planning and modelling is essential.

It is anticipated that planning will be site specific and propagation models must accurately predict power levels as well as temporal and spatial channel characteristics. For example, the angle of arrival information of each of the multipath radio wave components can be used to estimate the capacity enhancement possible from advanced technologies such as beam-steerable antenna systems.

Current propagation modelling techniques are based on ray-tracing, but such systems must make assumptions. Accuracy is fundamentally limited by the level of detail available about the environment as well as the amount of computational power available to run the simulation. Currently, ray-tracing simulations result in poor accuracy when modelling deep shadow regions, commonly found in urban environments.

The VURI has designed and built a channel sounder to accurately measure the spatial and temporal channel characteristics of radio environments. The channel sounder hardware combined with computer post-processing allows construction of a 'picture' showing what the radio environment looks like at different time instants (see Fig 1.A1 for an example).

The results from such studies should improve the accuracy of planning tools and simulation software. It is hoped that propagation modelling techniques will improve sufficiently to allow accurate models to be made despite fundamental limitations. However, it is unlikely that optimum network performance will be achieved without performing some measurements at each cell site.

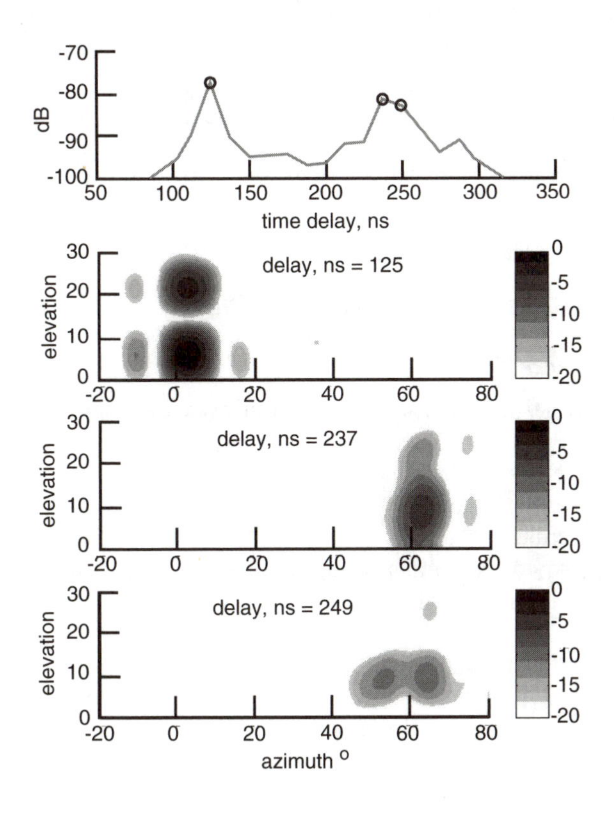

Fig 1A.1 Channel impulse response (CIR), shown top, demonstrates the variation in signal strength against time. Three radio images (bottom) show direction of arrival information for the peaks marked on the CIR. The data was collected at Birmingham University and forms part of a paper [8] at VTC2000 [9].

Several VURI papers have been produced [4] through collaboration between the various VURI members and BT researchers. Further work is under way, including construction of a more accurate and portable channel sounder.

A BT joint venture company has already expressed interest in buying BT's channel sounding equipment, and more interest is expected as the industry struggles with the current models and planning techniques. The aim is to ensure BT Wireless has this technology available — ahead of it competitors — and before it becomes essential.

References

1 Communications Engineering Group, Birmingham University — http://www.eee.bham.ac.uk/com_gr/

2 Centre for Communications Research, Bristol University — http://www.een.bris.ac.uk/Research/research_index.htm

3 Engineering Science Department, Oxford University — http://www.eng.ox.ac.uk/~comwww/

4 VURI Web site — http://www.btvuri.com/

5 SDR Forum Web site — http://www.mmitsforum.org/

6 Mobile VCE — http://www.mobilevce.co.uk/

7 Mitola, J.: '*The software radio architecture*', IEEE Communications Magazine, pp 26-38 (May1995).

8 Mughal, M. J., Street, A. M., and Constantinou, C. C.: '*Detailed radio imaging of buildings at 2.4 GHz*', IEEE Vehicular Technology Conference (September 2000).

9 IEEE Vehicular Technology Conference 2000 — http://www.vtc2000.org/

2

VIRTUAL CENTRE OF EXCELLENCE IN MOBILE AND PERSONAL COMMUNICATIONS

K A Holley and K G Hall

2.1 Introduction

During the 1990s the UK government had a successful series of programmes, entitled LINK, which led on to an initiative called Foresight. This initiative brought together industry, the academic community and the government to identify where future academic research should be focused in order to serve the needs of wealth creation. The Foresight Challenge highlighted the importance of long-term leading-edge academic research into mobile and personal communications. This led to the creation of the Mobile Virtual Centre of Excellence[1] as a university/industry partnership, aimed at preserving the UK's world lead in mobile communications and to secure the future wealth created from it. The objectives were defined at the outset as:

- to harness the research efforts of a selected group of UK universities into a cohesive, world-class virtual (geographically distributed) research centre of excellence, focused on mobile radio and personal communications;
- to provide a framework for industry to fund research in UK higher education establishments;
- to influence the direction of long-term research through a continuation of the Foresight process involving dialogue between industry and the academic community;
- to carry out programmes of research in a manner that makes best use of the academic creative talent existing in UK universities, with particular emphasis on encouraging a cross-disciplinary approach and well-managed programmes;

[1] The executive director of the Mobile VCE, Dr Walter Tuttlebee, kindly provided assistance with, and provision of key material for, this chapter. Some of this material was presented at the IEE 3G 2000 conference; we gratefully acknowledge the IEE's permission to reuse this material.

- to provide a mechanism for industry to work collectively with key universities to ensure it has the necessary flow of the most skilled experts in the new technologies;
- to project its work internationally in order to attract research funds and to sustain the world-wide reputation of UK universities for innovation.

The rationale has been to harness a partnership between industry and academia to work on a longer-term research programme for the benefit of the UK. The setting up of Mobile VCE, while not without considerable challenges, has been achieved in a timely and efficient manner resulting in a highly flexible entity. This has been achieved in no small part by the outstanding response to the initiative by the mobile communications industry, resulting in a membership which is truly representative of the industry not only in the UK, but globally. The result is a strong organisation which yields outstanding gearing of the funds received.

2.2 Objectives

The prime objective of the Mobile VCE is to undertake long-term research and, within a clear framework, to create commercially valuable IPR relevant to mobile/ wireless communications for its members through high levels of active involvement. The large scale of the research programme, over £1m per annum, and the commitment of the major industry players has created a momentum that has encouraged the participation of senior technical staff from the mobile telecommunications companies. This will be a key factor in successfully downstreaming the research into industry developments.

The Mobile VCE delivers its results by means of a series of core programmes, each with its own objectives. For Core 1, which was completed during the year 2000, the objective was to set up and establish a key skill base of knowledge within the universities, and to deliver some initial ideas and patents. For Core 2, the objective is to provide high-value solutions for industry, starting from third generation mobile networks and identifying ways in which technology can be grown from there.

2.3 Membership

Industrial members may join on payment of an annual membership fee and acceptance of the memorandum and articles of association of the company, and upon signing a deed of adherence to the company's IPR agreement. Currently Mobile VCE has 21 such members — leading international players in the wireless industry (see Table 2.1) who each pay an annual subscription of around £30k or £60k, depending on whether they opt for a Class 1 or Class 2 membership.

Membership is viewed as a long-term commitment, at least for the duration of a Core Research Programme (3-4 years), rather than simply as a one-year decision. The financial costs, even taken over a 4-year period, are remarkably low compared to the benefits.

Table 2.1 Full industrial members of Mobile VCE, as at summer 2000.

BT Cellnet	Dolphin	Ericsson
Fujitsu	Hutchison 3G	Independent Television Commission
Inmarsat	Lucent	Motorola
NEC	Nokia	Nortel
NTL	One-2-One	Orange
Panasonic	Philips	Racal
Radiocomms Agency	Simoco	Texas Instruments
Toshiba	Vodafone Airtouch	

To encourage SME involvement a category of industrial associate membership exists, offering a slightly reduced set of member privileges in return for a smaller annual subscription, based on a sliding scale according to the size of the organisation.

Mobile VCE delivers value to its industrial members in many different ways, the main ones being:

- industrially relevant, long-term research — defined and steered by the members;
- highly cost effective — due to the significant financial gearing of members' subscriptions (which was ~40 times for the Core 1 programme — see Fig 2.1);
- intellectual property — the existing, already significant, IPR portfolio will increase substantially in future years with royalty-free licences available to full industrial members;
- short-term technology spin-offs — structures in place for commercial exploitation;
- elective research — funded by a single company, or by a small group of members on a shared-cost, shared-benefit, basis, to address shorter term R&D needs;
- training of own staff — through working with highly competent researchers;
- recruitment — development of new research staff for the industry and access to identify and recruit the leading research talent available today in the UK;
- networking — to identify and shape the trends which will influence the short term, as well as the long term.

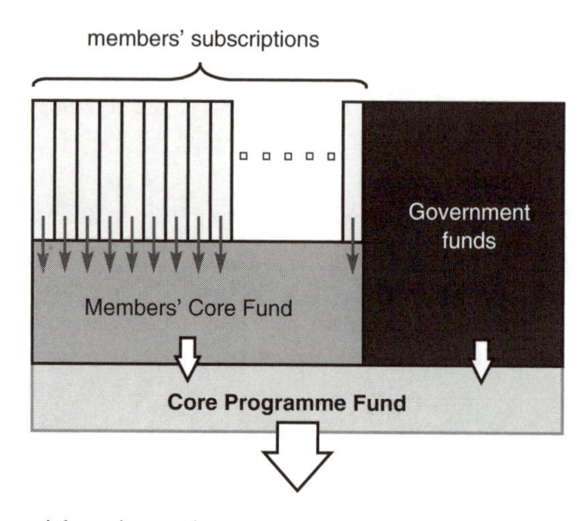

Fig 2.1 High financial gearing — the core research programme is funded by the aggregation of all the members' annual subscriptions, further augmented by Government funds.

As noted earlier, the Mobile VCE ties its research into industrial reality — the contribution of the industry staff, while low in terms of effort per company per year, is crucial in shaping the direction of the research. Different companies choose to put in different levels of manpower, with those who play a more active role deriving proportionally greater benefits.

For BT, there are a number of attractions in being involved in the Mobile VCE. The general benefits were outlined in the previous paragraph. BT has taken a 'hands-on' approach to the Mobile VCE and has involved a number of employees in the steering committees. Being involved at this level ensures that the mobile research work is relevant to the company and current strategic thinking, and that it is not overlapping with work being done elsewhere, for example in the various standards forums with which BT is involved. The Mobile VCE is also a member of ETSI in its own right, which also helps to avoid overlap with ongoing standards development.

The Mobile VCE has responsibility for assuring the quality of work performed by the university departments for the core research programme. This is achieved by only allowing universities to join by invitation, on the basis of assessment by the industrial members. One of the key criteria is a proven and clearly demonstrable track record — not just potential — of excellence in relevant research. When Mobile VCE was formed, some 23 universities applied for membership, of which just 5 were chosen; a further 3 universities have subsequently been admitted (see Table 2.2). Ongoing performance, assessed through formal procedures and by peer review by the industrial members, determines the level and areas of participation in future research programmes, thereby ensuring Mobile VCE maintains the highest quality, as befits a true centre of excellence.

Table 2.2 Academic members of Mobile VCE.

Bradford	Bristol	Edinburgh
Kings College London	Royal Holloway College	Strathclyde
Southampton	Surrey	

2.4 Steering Groups

Industrial members have full access to the Mobile VCE core research programme, technical reports and research teams, as well as the opportunity to participate in the steering groups, established for each of the core programme work areas. Steering groups comprise some 6-8 industrial members who meet quarterly to monitor, review and direct the research activities. Each steering group is chaired by a senior industrial member, supported by an academic co-ordinator who oversees the research teams.

The appointment of senior level industry technologists, often from the operational as well as the research parts of the industrial companies, to chair and participate in these groups enables a very strong input to the academics. It secures a higher degree of industrial relevance than is achieved in most other models of collaborative research. The structure and ownership of Mobile VCE ensures, however, that such industrial relevance can be achieved without compromising either the long-term nature of the research or its academic integrity.

This kind of involvement demands a high degree of commitment, but it does mean that for those who take the time, it can be ensured that the research work is relevant, correct and does not overlap with other activities in which BT is engaged. Through this commitment, early knowledge of research results can also be gained. Close working within the teams can result in the movement of academic staff into industry roles for shorter or longer durations.

2.5 Pan-University Teams

The research within each work area is undertaken by collaborative teams, usually comprising personnel from more than one university. Co-ordination of these pan-university teams is managed by an academic co-ordinator, a senior staff member — lecturer or professor — from one of the member universities. Innovative 'carrot and stick' mechanisms have been established within Mobile VCE that have proved to be very effective in encouraging the universities to deploy high-quality creative staff within the Mobile VCE teams, to collaborate closely across traditional boundaries and to identify valuable IPR. Joint pan-university publications have begun to appear, as have complex software simulations tools, which have been created in a composite manner across multiple research institutions. When Mobile VCE was first established such collaborative achievements were just hopes, but have now become reality.

2.6 Research Programmes

Two types of research activity are undertaken by Mobile VCE — integrated so-called 'core' research programmes and elective research. Both types of programme (described below) are defined jointly by industrial members and leading academics, are monitored and guided by steering groups, and are undertaken at member university premises, usually by combined teams from more than one university, selected to match expertise to requirement. The research is based on formal contracts placed by Mobile VCE with its academic members, and is overseen by the Mobile VCE office, which provides a central hub for management, communication and information dissemination

Core research programmes represent substantial integrated programmes of research, funded by pooling members' subscriptions, usually also augmented by Government grants.

The first core research programme, some 56 man-years of focused research, commenced in March 1997 and concluded in early 2000. The programme comprised four work areas:

- services;
- networks;
- terminals;
- radio environment.

The Core 1 programme delivered over 20 patent filings, of which 11 have been pursued to the PCT (international) stage. A large number of detailed research reports, provided on CD-ROM to members, are complemented by members' technical seminars, covering each work area and held each year. Such seminars provide opportunities for a much broader base of technical staff from member companies to become aware of the results and resources available to them through Mobile VCE. Detailed data and processed results from advanced propagation research campaigns are also available to members on CD-ROM, as are software tools relating to multimedia traffic planning and other topics.

A new core research programme, Core 2, recently commenced, the objectives of which relate to a year 2010 scenario. In all probability there will be shorter term spin-offs, as there were from Core 1, where research contributed to and influenced companies' in-house programmes and thereby fed indirectly into standardisation.

The Core 2 research programme, described on the Mobile VCE Web site [1], received a favourable critical evaluation from a panel of international experts, from both industry and academia, who reviewed it as part of the EPSRCi decision to support it with grant funding. It is a major research programme, of over 100 man-years, structured into three work areas:

- software-based systems;
- networks and services;
- wireless access.

The enlargement of the programme, compared to Core 1, has enabled Mobile VCE to widen its academic base, embracing an increased emphasis on mobile computing, middleware, agent technology and security aspects, complementing, and building upon, existing strengths in wireless access, software radio, networks, and services

Mobile VCE also undertakes 'elective programmes' — research programmes of interest to perhaps just a subset of the industrial membership and in which members may choose ('elect') or decline to participate. Such programmes are funded by either a single company or jointly by a group of companies with a common interest. For such programmes, access to results and any consequent IPR is restricted to those who elect to fund the work. Such programmes may build, for example, on prior core research, providing members with an easy route for cost-effective pre-development of emerging technology, bridging the gap between research and product development and ensuring that the companies have direct access to the relevant researchers, easing the transition from research to product.

The elective mechanism also provides a route for industrial members to outsource research work for which they lack sufficient specialist internal staff, using Mobile VCE as a flexible extension resource to their company. Such work is obviously undertaken on terms of strict commercial confidentiality.

2.7 Publications

The Mobile VCE has produced a number of papers in the past, information about which is available on the Web site [1]. Additional information on the research achievements is available in a special issue of the IEE Electronics and Communication Engineering Journal [2].

2.8 Conclusions

The Virtual Centre of Excellence in Mobile and Personal Communications — Mobile VCE — is now some four years old. Through the Mobile VCE, BT has access to a wide range of academic expertise which it is able to use to focus on the challenges of taking mobile technology to the next level, without having any overlap with its other activities, such as 3G standards work, and the VURI (see

Chapter 1). The Mobile VCE has also developed a wide patent portfolio, to which BT has royalty-free access.

Using the work of the Mobile VCE, therefore, enables BT to put together new technology for future deployment within its mobile networks, gaining significantly both from the different perspective that academia has to offer, and from the high gearing provided through being members of a large industry community.

References

1 Virtual Centre of Excellence in Mobile and Personal Communications Web site — http://www.mobilevce.co.uk/

2 IEE Electronics and Communication Eng J, Special issue (December 2000).

3

3G PRODUCTS — WHAT WILL THE TECHNOLOGY ENABLE?

J A Harmer and C D Friel

3.1 Introduction — Evolution from 2G to 3G

Mobile telephones have been a tremendous success story. Although GSM was originally conceived as a pan-European system there are now over 400 operators in 169 countries worldwide [1]. Since the launch of the first GSM systems in the early 1990s, customer numbers have grown very quickly resulting in today's highly competitive mass market. In particular, the more recent development of 'pre-pay' packages has been very successful in attracting new customers. For many mobile operators more than 50% of their customers use pre-pay. At present, GSM use is dominated by voice services, although there has been a great increase in uses for the short message service (SMS), a text-based messaging system. GSM provides a 'circuit-switched' data service similar to PSTN dial-up data services. The nominal GSM data rate is 9.6 kbit/s. In recent years, a new coding scheme has been approved that takes this data rate to 14.4 kbit/s although not all operators will adopt this. However, these data rates do not provide high-speed access to services such as e-mail and the World Wide Web (WWW). Also, the use of 'circuit-switched' connections, where the channel is dedicated to one user, is not the most efficient way of carrying the 'bursty' traffic of these types of services that are accessed over the Internet.

Why evolve to 3G? Just as the advent of the WWW has generated a 'datawave' on the fixed network, it is expected that the same will happen to the mobile network and that there will be a move to using data services on mobile devices. The rise of fixed network Internet access has led to global connectivity. E-mail communication has been a very strong driver. Until now the Internet and mobile communications have grown separately. A challenge for 3G is to bring the best features of mobile communications and the Internet together. Different markets may evolve at different rates, e.g. Europe has medium-level fixed Internet penetration but leads the USA with high mobile communications penetration (based on the one standard — GSM). Conversely, the USA has very high Internet penetration, but an array of different mobile standards and less complete national coverage. Japan currently leads the way with mobile eCommerce applications and services through i-mode services available via NTT DoCoMo's PDC packet data network [2].

2G is evolving from a voice-centric starting position, and enhanced 2G technology is being rolled out to offer more advanced data type services; the general packet radio service (GPRS) development of GSM enables mobile Internet. 3G networks are poised to enable mobile multimedia.

3.2 Will Mobile Data be a Success?

Much analyst attention has been focused on producing forecasts for the growth of 3G networks and the mix of voice and data traffic. On GSM, voice-centric services remain dominant but there is currently very high growth in the use of text messaging (SMS). For example, in the UK there has been a tenfold increase in the use of SMS over the last 12 months, half a billion messages per month in the UK alone, from July 2000 [3]. The user benefits of SMS are low cost, alerting and notification capabilities, and an off-line and non-intrusive means of communication. The market for mobile data is already growing. BT Cellnet's Genie Internet [4] is now providing access to content over wireless application protocol (WAP) telephones.

The evolution of products is tied to the capabilities of terminals and the network, e.g. it is possible to demonstrate video over GSM circuit-switched data now, but the performance is very limited.

At launch, 3G networks will carry a very high proportion of voice traffic. By 2010 it is predicted that more than half the 3G network traffic will be data related. Most of this will be WWW and e-mail type services. The highest quality and most demanding interactive services (such as real-time video, high-quality audio and gaming) are likely to represent no more than 10 per cent of total network traffic. Given the recent upsurge in the use of SMS, it is possible that data traffic could also increase much more quickly than anticipated.

3.3 The 3G Product Value Chain

The 2G GSM value chain (Fig 3.1) is relatively straightforward. Network operators (NOs), service providers (SPs) and value-added service providers (VASPs) are the main players. A user's choice of products is restricted to what is offered by the NO or the SP from which they bought the service (e.g. BTCellnet 'Find Me' and 'Traffic Line' products).

Access to Internet-based services from mobile devices introduces a more complicated value chain. In order to appreciate the significance of products in 3G business it is important to understand the value chain (Fig 3.2) and sources of revenue.

Fig 3.1 2G value chain.

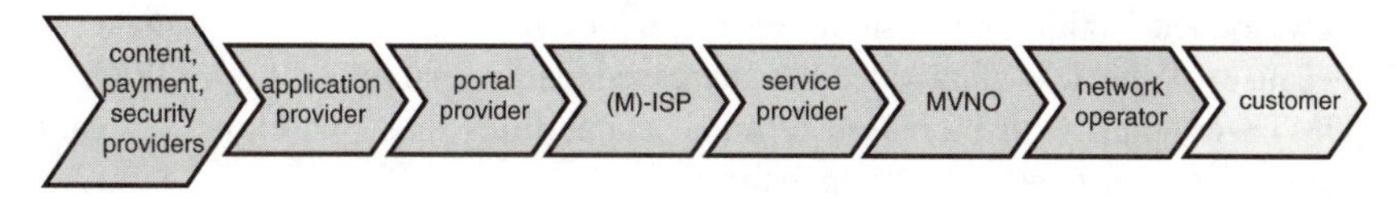

Fig 3.2 3G value chain.

- Network operator

  NOs have a mobile licence and own the radio spectrum, base sites and switching centres. Their function is to provide network access ('conveyance'). Sources of revenue are through retail via their own service providers, and wholesale airtime to other service providers and mobile virtual network operators.

- Service provider

  SPs buy wholesale airtime from NOs and package and sell it on to retail customers. SPs issue SIM cards and own the billing relationship with their customers.

- Mobile virtual network operator (MVNO)

  MVNOs usually own more network infrastructure than SPs. It is possible that they could own the complete switching infrastructure — everything except the radio interface. MVNOs have started to appear, e.g. Virgin Mobile [5] in the UK. MVNOs are also proving to be innovative, for example, by offering new services using the SIM application tool-kit (a capability that can customise the mobile telephone menu to provide additional services). Although there are only a few MVNOs now, numbers are expected to increase dramatically with 3G. In a few years, there could be thousands of virtual operators owning SIMs and customers, and buying wholesale airtime from NOs.

- Mobile Internet service provider (M-ISP)

  M-ISPs terminate data calls on an IP network and provide users with IP addresses and authentication.

- Portal (context) provider

  Portal providers give a mobile homepage and set of services for mobile users. Revenue is generated by the 'stickiness' of the site — advertising and referrals lead to revenue from content providers if users choose to stay with the site as their chief source of information and search capability. The freedom of mobile users to select which WAP portal they wish to access from their mobile telephone has been much publicised.

- Application provider

  Application providers supply products which are purchased by users either by pay-per-use or subscription. Applications are either used on-line or are downloaded. While not usable over the circuit-switched GSM data service because of the low data rate, the increased data capacity of both GPRS and UMTS coupled with increased terminal capabilities make wireless application service provision much more appealing.

- Content provider

  Content providers are powerful players in the Internet and mobile Internet applications space. They sell or license their content to portal providers.

- Payment processing provider

  The payment processing provider may either support individual products or wholesale payment processing to other companies, content providers or service providers.

- System security provider

  System security providers are likely to increase in significance as mobile access to corporate information becomes widespread, and financial transactions become possible from mobile devices.

3.4 The 3G Product Revenue Challenge

The mobile Internet value chain parallels that for fixed Internet services. Data services are set to move from being circuit-switched to 'always on' and therefore priced at a flat rate. In some countries the number of 3G licences awarded is greater than the number of incumbent 2G operators, creating new entrants. Coupled with the opportunities for service providers and MVNOs, this means that there will be major challenges for 3G licence owners.

In Europe, 3G licences have cost the industry €150 billion so far. Over time, mobile users will expect significant reductions in conveyance costs (along the lines of the flat rate, fixed Internet model, e.g. ADSL). 3G network operators will need to generate revenue from other sources by:

- being active in multiple layers of the value chain — retail, wholesale, and content and application hosting;
- partnering with best-of-breed content and application providers;

- generating revenue from selling value in their network, e.g. location, alerting;
- wholesaling components to third parties, e.g. billing, payment processing, SIM card issuing, product configuration and management.

This means competing for business with wired and wireless service providers, portal providers and content providers. Compelling products will be the key to success.

3.5 3G Products at Launch — What Will the User Benefits Be?

The role of 3G products will be to provide customers with innovative, high-quality applications for which they will be willing to pay. Although network access costs are expected to fall considerably, there will still be a premium for mobility for which customers will be prepared to pay, as they do today. However, most customers will be sensitive to the price of the product to varying degrees. The advent of UMTS will give network operators the opportunity to clearly demonstrate the superior capabilities of 3G technology. This will be done by the deployment of new products as well as the enhancement of products used on 2G systems.

So what is a product? A product can be defined as a package of specific capabilities of the network and associated application and service platforms designed to provide the user with a useful function, e.g. a 'mobile office' product could offer the customer access to corporate e-mail, diary and scheduling services, while also offering high-speed access to the corporate LAN and voice access to the company PBX. To do this, the mobile office product will need to use a combination of high-speed data access, real-time voice capability and the use of appropriate service platforms, such as multimedia messaging, security and alerting.

Products will be offered by both the network operator and third parties. Through open interfaces and wholesale agreements, UMTS will make it much easier for third parties to package the basic network capabilities and connect their own platforms. There will therefore be much more competition in the market-place and the user will be faced with much more choice.

The basic features of UMTS that will enable this competitive market are high-speed data access, efficient multimedia handling, and standardised interfaces to allow third parties to create their own products. In the future, UMTS will be more closely aligned with the Internet by using IP technology and users will therefore benefit from even more products that can be delivered to users by both fixed and mobile access in a seamless way.

The following sections look specifically at how 3G technology will enable products which bring benefits to users in business and consumer segments.

3.5.1 Business Use

Increasingly, communication by mobile telephone is critical for large and small companies alike. 3G networks will enable good-quality voice products — the W-CDMA technology used for 3G is expected to give at least as good and probably better voice quality than GSM. This is expected to improve further with the advent of wideband speech codecs. For voice services, the existing GSM supplementary services, such as call forwarding and barring, will continue to be available. Such services, however, will be enhanced to offer the user control of multimedia, e.g. control of voice and video independently.

The increased capacity and bandwidth of UMTS will provide the business user with many additional multimedia benefits:

- video applications, such as videotelephony and conferencing;
- fast access to corporate systems, such as intranets and mail systems;
- multimedia messaging and rich content, including graphics for information services such as directory listings, travel information and schedules.

Location-based services, such as fleet tracking or work management, will enable more efficient use of business resources. More powerful terminal devices with full multimedia capability will allow the user to have a fully interactive session with the services and products using human-friendly interfaces such as speech recognition and voice simulation. Another important requirement for the business user will be the ability to synchronise many different devices. For example, the synchronisation of personal organisers with information held on business-use PCs to ensure consistent schedule and contact information. This could be realised by a combination of network intelligence and device connectivity (e.g. by using Bluetooth [6]).

Similarly, a business will require the ability to manage the configuration of their corporate devices. This will require the provision of automated processes and flexibility to allow this to be done either by the corporate customer directly or by a third party on behalf of the corporate customer.

3.5.2 Consumer Use

2G networks and terminal devices offer only basic entertainment products. For example, some devices offer stand-alone games (e.g. Snake, Tetris) with, in some cases, the ability for two or more players to interact using the infra-red port on the user device.

Forms of entertainment that will be possible over 3G systems include the delivery of live output from radio stations, film trailers and sampling of the latest popular music. Audio- and video-streaming capability will be used to deliver these products, which will be developed and enhanced as the capabilities of 3G systems improve — particularly as data rates increase.

Localised services can be provided to the customer depending on their location. Some of these services (such as local traffic reports) are already provided on 2G networks. The greater capability of 3G will allow more sophisticated services to be provided and users can also expect to receive multimedia advertisements on their mobile from local businesses.

With the advent of 3G and advanced consumer devices, the higher bandwidth could even allow real-time interactive gaming with multimedia presentation.

3.6 3G Product Development — More Capabilities after Launch

Further releases of the UMTS system will give enhanced capabilities. Consequently, products will also develop and improve. One of the main goals of UMTS development is to base all the architecture on IP technology. This is the same technology that is currently used by the Internet, but it will need to be enhanced to allow the support of real-time services such as voice and user-to-user video. The move to IP technology will allow full integration of 3G-based mobility and the Internet. This will bring benefits such as the ability for users to seamlessly access products and services from both the fixed and mobile networks. In fact, it is predicted that only a few years after 3G launch, more people will access the Internet from mobile devices than fixed PCs. This will mean that users will (if they want to) have access to Internet-based services and products, e-mail and other messaging systems at all times, wherever they are.

Products will evolve to allow users to have complete flexibility of access. It is expected that there will be a significant increase in the 'localisation' of services since the user's position can be easily determined. Positioning technologies are reviewed in the following section.

3.7 3G Product Enablers — a Component-Based Approach

3G products inherit much from Internet and multimedia products rather than from traditional voice services. This is supported by the future realisation of 3G based on IP technology. For these reasons, the 3G standards propose a decoupling of features and capabilities from products and applications, leaving a great deal of opportunity and freedom for third party product developers. 2G products are highly standardised and are tightly coupled with the network operator's network. In contrast, 3G products will be offered not only by the network operator, but also by many other third parties. In both cases, the products will be built from the basic capabilities of the network. These include network capabilities (e.g. provision of a bearer, quality of service, supplementary services) and the service platform capabilities (e.g. messaging, location servers).

One of the features of 3G is the provision of an open service architecture and service creation environments that will allow third parties to access network capabilities and service platforms in a standardised way in order to build products. In the same way that open Internet protocols and interfaces have enabled a wealth of IP-based products, open interfaces to 3G will enable a wide range of innovative products. The open service architecture and the service creation environment are shown in Fig 3.3.

A number of key components can be defined which are essential for 3G products.

3.7.1 Personalisation

Mobile users do not want to spend time searching for information. Profiling technology can be used to match content (e.g. news, stock market information) to the needs of individual users. Profiling also means that less user input is necessary on small devices which generally have limited input capability (e.g. stylus, on-screen keyboard).

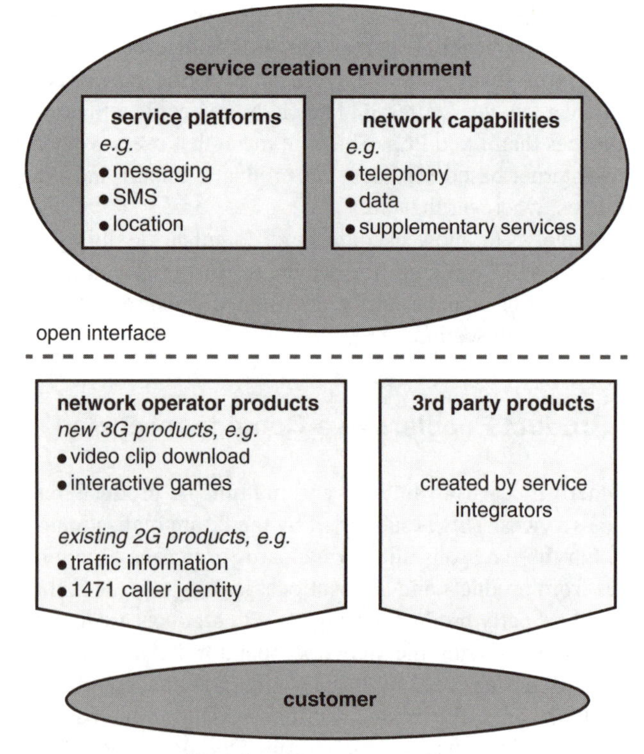

Fig 3.3 3G service creation.

3.7.2 Immediacy

GSM SMS has been highly successful as a means of alerting users to new information, e.g. share price, new mail, football scores. 3G will extend this capability to include pictures, audio clips and video clips as part of the multimedia messaging service (MMS). Users will therefore benefit from much richer information than a 160-character SMS can provide today.

3.7.3 Quality of Service (QoS)

QoS is key to the quality of the user experience. There is a trade-off between data rate offered, the cost of service and user experience. A particular concern for operators is how to offer multimedia applications while ensuring that acceptable end-to-end QoS can be maintained and delay minimised. Network operators and their partners and alliances have the ability to provide QoS-controlled backbones. A challenge is to provide consistent-quality roaming access for packet-based services at a price that is acceptable to network operators, service providers and customers. It is also critical for network operators to meet QoS obligations to service and application providers.

3.7.4 Ubiquitous Access — Roaming and Handover

As with the existing GSM standards, a key feature of UMTS will be the ability to roam. The ideal requirement would be for the user to be able to take their handset anywhere in the world and use it to access all the products and services enjoyed in their home country. In practice, this will not be possible and limitations will exist.

One of the main problems, certainly in the first few years, will be lack of complete 3G coverage. Many countries are committed to the deployment of 3G systems but many will not have 3G systems in full service for many years. Generally, it is expected that much of Western Europe and Japan will be covered by 2005 (driven by licence obligations), but significant holes will exist in important areas, particularly in the USA.

The situation in the USA is complicated by the number of different types of 2G system in use and also by the use of some of the 3G designated spectrum by existing 2G systems.

It is likely that many US operators will deploy enhancements to their existing 2G networks first to provide them with an 'almost 3G' network. In the case of US GSM operators, it is likely that enhanced data rates for GSM evolution (EDGE) technology will be used. US operators who do not use GSM will choose a different upgrade path and will probably finally deploy the main 3G alternative to UMTS that is known as CDMA2000.

For many years, it is likely that the patchy UMTS coverage will be supported by GSM/GPRS, both nationally and internationally. For this reason, it will be important that UMTS terminals are also capable of using these existing 2G networks. This will allow the user to roam on to 2G networks where there is no UMTS. It will also allow the user to handover between 2G and 3G networks.

The 2G/3G handover capability gives continuity of service for users as they move in and out of 3G coverage in a given country or area. The call or data session is continued on the new network without users having to re-establish the call or session. Ideally, users should not notice this changeover of network but in many cases handover from 3G to 2G will cause a noticeable degradation.

The main problem is that the 2G network will not be capable of supporting the data rates required by the product or service in use. This should not be noticeable for voice but will be significant for large bandwidth services such as live video. For this reason, it is important that the handover process can degrade the product performance gracefully. An example is a videophone call. A 2G network will not be able to handle both the voice and the video components of the call because of the large data rates required. In this case, on handover from 3G to 2G, it would be desirable for the voice component to continue and the video component to be dropped. The video component would be re-instated when the call is handed back to the 3G network. In practice, this may prove difficult in the early years of UMTS and some other means of managing the handover of this type of product will need to be determined.

3.7.5 Location-Based Services

The ability to locate a mobile user can enable a variety of applications ranging from emergency call to locating people or places. Much work has been done in the 3GPP groups to develop standards for location. This has been driven from the USA where the FCC E911 initiative requires mobile operators to be able to locate users making emergency calls with an accuracy of 125 metres or better, 67% of the time. This capability is required by October 2001. The 3GPP standards identify a number of location fixing systems (LFSs). Their main approaches are as follows.

- Cell of origin (COO)

  Most mobile location-based services available today use positioning based on the cell of origin (from which the call was made). This can be combined with signal-timing measurements to enhance the accuracy of the positional fix. As overall accuracy depends largely on the sizes of cells, performance would be better in urban areas than in rural areas. In urban areas, accuracy of better than 150 metres can be achieved. No modifications are needed to the mobile terminal or the network for the COO-only solution (minor modifications are required to incorporate timing measurements) and the technology is available now. This technique can be used on both 2G GSM and 3G UMTS systems.

- Assisted GPS

  The global positioning system (GPS) is a satellite-based system using a constellation of 24 satellites. These satellites constantly broadcast their own timing information. A GPS receiver is used to receive the signals from a number of satellites, which enables the position of the receiver to be calculated. However, GPS requires line of sight to the satellites in order to receive a signal, and does not work within buildings and in 'urban canyons'. Assisted GPS is currently being developed for mobile telephones. The mobile device is given information about which GPS satellites are in view. This helps to reduce the processing power that is needed, and gives a faster start-up time and improved sensitivity. An assisted GPS solution is being developed by SnapTrack and trials have been undertaken in Madrid and Paris.

- Time of arrival (TOA)

  TOA does not require any modification to mobile handsets. The timing information in the network is used to calculate the position of the mobile. This timing information must be received by three location management units (LMUs) in the network. Large numbers of LMUs are required in the network which make TOA a relatively expensive solution. Accuracy ranges from 50 to 150 metres. The need to use handovers to get the measurements means that the technique will not scale well to large numbers of location requests. This technique can be used on GSM systems but is not suitable for UMTS which uses an incompatible type of radio interface.

- Enhanced observed time difference (E-OTD)

  Time-of-arrival signals from base-stations are measured by mobiles and also a location-measurement unit in the mobile network. Information from three base-stations can be used to resolve the location of the mobile. An LMU is needed for every 3-5 base-stations. Ericsson [7], Nokia [8], and Cambridge Positioning Systems [9] offer E-OTD solutions. This is the 2G technique — on 3G systems (UMTS) a similar technique is used that is known as observed time difference of arrival (O-TDOA).

- Provision to third parties

  Currently the location of mobile users is known only to cellular network operators. In the future it will be possible to provide location information on a commercial basis to third party product developers and application providers. However, a framework for maintaining user rights to privacy and anonymity must be put in place. In any case, it will probably be mandated that a mobile's location be provided to certain third parties (e.g. emergency services).

3.7.6 Security

Developments in mobile telephone SIM cards and convergence with the smartcard industry could lead to potential for using the mobile device as a security token. Already, trials with dual-slot GSM telephones have shown that products can integrate GSM security with a smartcard eWallet issued by a bank.

3.8 Mobile Terminals are Key to the Customer Experience

For UMTS, the terminal manufacturers are producing four main types of device that will suit different types of product and application:

- voice-centric — mostly for the use of voice services but may have some simple additional capabilities as well;
- PDA — personal digital assistant, which will combine a mobile telephone with personal organiser-type functions;
- smart phone — voice-centric type with some additional processing capability to enable facilities such as video transmission, eCommerce transactions, etc;
- data card — a device that will plug into a PC/laptop mostly to handle data transmissions, e.g. Internet and e-mail access.

The use of local connectivity technologies such as Bluetooth will allow different components to be fitted together in a local area to configure a device as the customer requires. From the experience of 2G, it is clear that devices are without doubt a fashion item for youth. For all segments they are a vital part of the customer experience and their performance is critical to ensuring that the customer enjoys good service from the network. Perhaps just as important as the technical capability and performance of the device is its appearance. Variety, colour and shape are key to marketing the device to certain sectors, particularly the youth market. Strong opportunities exist to sell the products provided by a network on the consumer appeal and functionality of the terminal devices.

A critical factor for mobile products is to enable content to be delivered to a wide range of devices with differing functionality over networks with varying capacities and capabilities. In the 2G arena there is currently a lively debate on the virtues of WAP versus cHTML versus HTML. When looking ahead to the increased capabilities of future terminals and networks, WAP is not the ultimate solution, but an important starting point. Dual stack (WAP/cHTML) devices are emerging now which give flexibility to operate with either presentation mode. XML technology provides the means to deliver content in an appropriate way to a very wide range of terminal devices. Nevertheless optimisation technologies are still a highly valuable way of improving performance and effectively using capacity over a wireless network.

3.9 Paying for 3G Products

Pre-pay has become a significant way of paying for 2G products. About 80% of new additions are pre-pay for operators offering the service. For many operators, this means that at least half the customer base comprises pre-pay customers. It should be noted that average revenue per user (ARPU) is considerably lower for pre-pay — about one third of the post-pay ARPU.

Technology developments in mobile terminals, SIM and smartcards lead to potential for 'eWallet' payment applications combined with pre-pay. Users will find it convenient to pay for items with their mobile, using a PIN for extra security. Already banks are offering mobile finance applications via WAP telephones. Alternatively, charges for products could be added to post-pay bills, or charged via a credit-clearing capability. Opportunities exist for 3G operators through wholesale operations, such as billing, payment and security, to third party product developers. Highly flexible billing and charging systems will be necessary to accommodate a wide range of billing and charging scenarios:

- by the minute (premium rate, e.g. videotelephony, high-speed data);
- by volume (packet data services, volumes bundled with subscription);
- per access to content;
- by purchase of goods;
- volume discounts, affinity marketing deals, special offers;
- advertising and sponsorship.

3.10 Customer Care

Just as the value chain becomes more complicated with mobile Internet and 3G, so will customer care. International and 2G/3G roaming increases the complexity further. When things go wrong it is essential that there is a reliable source of information to enable customers to look after themselves as much as possible. WWW enables on-line provisioning. Technologies such as over-the-air programming, with security enhancements, or Java downloads could enable remote configuration and maintenance.

Today's WAP telephones require configuration and there are already many different browsers. Future 3G devices will be more diverse and will require simple set-up either 'out of the box', or by a single call/click. Easy-to-use customer self-care should be a priority to avoid prohibitively high costs of set-up and maintenance and to avoid customer frustration and disappointment.

3.11 Conclusions

The technology offered by 3G will enable products to offer much more than just mobile access to the Internet. Network operators and application providers need to exploit the features of 3G networks which add value and enable revenue to be gained from applications and service provision in addition to conveyance:

- personalisation — the right information;
- immediacy — the right information, exactly when it is needed;
- QoS — the right value to match the product;
- roaming — wherever you are;
- location — information relevant to where you are;
- security — for you alone.

These components and others, accessible via open interfaces, will be available to a large community of application developers. It is vital that this pool of innovation is encouraged and exploited in order to develop the market for 3G products and generate revenue for all parts of the value chain.

References

1 GSM Association Web site — http://www.gsmworld.com/gsminfo/gsminfo.htm

2 NTT DoCoMo Web site — http://www.nttdocomo.com/i/index.html

3 MDA Web site — http://www.text.it/navibar/naviset.htm

4 Genie Web site — http://www.genie.co.uk/

5 Virgin Mobile Web site — http://www.virginmobile.com/mobile/

6 Bluetooth Web site — http://www.bluetooth.org/

7 Ericsson Web site — http://www.ericsson.com/

8 Nokia Web site — http://www.nokia.com/networks/mobile/

9 Cambridge Positioning Systems (CPS) Web site — http://www.cursor-system.co.uk/

4

THE DEVELOPMENT OF MOBILE IS CRITICALLY DEPENDENT ON STANDARDS

F Harrison and K A Holley

4.1 Mobile Standards Background

The importance of standardisation to the mobile industry is probably best illustrated by the impact of the work on the second generation mobile systems, in particular the GSM (Global System for Mobile Communications) standard. The work was started in the mid-1980s, at about the same time as the launch of commercial first generation cellular networks. A variety of different first generation systems were deployed across Europe and the rest of the world. Network infrastructure and mobile terminals were relatively expensive and market adoption was mainly limited to business users. The objective of GSM standardisation was to create a new pan-European system, where users could continue to get service when roaming across country borders. Economies of scale of larger production volumes were expected to reduce prices.

The impact of the GSM standards work is now clear to see. GSM is effectively a *de facto* world standard, claiming around 60% of the total mobile telephone population, and with coverage in all the continents. GSM terminals and network costs have reduced to such low levels that consumer pre-pay packages can be retailed at £50 or less, and the consumer market has developed such that the penetration of mobile telephones exceeds 50% of the population in many countries.

The mobile communications industry has grown to the point where GSM manufacturers, such as Ericsson, Nokia and Motorola, and global operators, such as Vodafone, are now among the top companies in financial market capitalisation.

The speed and extent of this development is undoubtedly due in large part to the creation of international standards. Had we continued by further developing the differing national first generation standards, the mobile industry could not have come so far, so quickly.

The challenge for third generation standards is to take this further — to create a truly single, world-wide standard. However, the standardisation environment has changed since the early work on GSM and it has been necessary to modify the approach. In particular, there are several key differences which have influenced the way in which 3G standards are being created:

- GSM had been created by the European Telecommunications Standards Institute (ETSI) as a European regional standard, and to create a world standard needs the involvement of other standards bodies, for example from the Far East and North America;
- the need to work on a world scale has introduced a wider variety of political and cultural influences on the standards work, where conflicting agendas can put the goal of a single standard at risk;
- substantial investments have now been made in second generation systems and infrastructure — whereas in the transition from first to second generation it had been possible to start with a clean sheet of paper, for third generation there is a strong commercial drive to reuse existing infrastructure and create an evolutionary approach, in order to capitalise on past investments;
- the extent of roaming capabilities with GSM is significant across the globe, and different networks will proceed at different rates towards third generation, but users of 3G will not accept a reduction in roaming capability and this requires intelligent handling of roaming between 3G and 2G — this situation will also lead to more complex combined 2G/3G handsets;
- the technology scope of 3G has broadened considerably to embrace data and Internet aspects — while GSM is a largely self-contained standard, for 3G it is necessary to draw on component standards from other bodies, such as the Internet Engineering Task Force (IETF) IP protocols, and this requires co-ordination and collaboration across different standards forums, where quite different standardisation processes are used.

These and other aspects led, in December 1998, to the creation of a new partnership project known as 3GPP (3rd Generation Partnership Project), involving several regional standards bodies working together.

4.2 Why Standards are Important

The development of standards benefits users and all sectors of the mobile industry — suppliers, operators and regulators. Examples of these are given in Table 4.1.

On the counter side, it is important that standards do not stifle innovation, and that product and service differentiation can still be achieved.

In practice, such is the importance of standardisation to the mobile industry that the creation of standards is a key part of the development process. New terminals and network systems simply cannot be launched without standards in place. One measure of the importance could be the scale of effort on 3G standardisation, where tens of thousands of engineers are already engaged in the standards process. At any time of any day, you could be reasonably certain that there is a mobile standards committee in session somewhere in the world.

Table 4.1 Benefits of standardisation.

Economies of scale	Operators benefit from lower cost infrastructure. Users benefit from low cost terminals. Suppliers benefit from large and stable market.
Competitive procurement	Open standards enable interworking of network elements from different suppliers, avoiding long-term lock-in to a single supplier.
Service portability	Users can change to different terminal equipment or move between service provider and network operators while retaining a similar service.
Roaming	Services can be made available outside the user's home network with the goal of having exactly the same look and feel.
Regulatory	Regulators can specify standards for spectrum use, giving a controlled, yet open and competiive basis for licensing. Operators and suppliers can access new markets without being disadvantaged by local standards requirements.
Simplicity of service	Services and terminals operate in a similar fashion, making it easier for users (e.g. use of + symbol for international calls).
Service interworking	Standards can be used to ensure that services work well across different networks.
Applications and services environment	Network and terminal standards enable the development of 3rd party applications, and allow the development of 3rd party value added services.
Spectrum efficiency	Standards enable radio technologies to work together, or in close proximity within the radio spectrum, without undue interference.
Radio technology harmonisation	Standards are needed to ensure that different technologies can work in harmony in the same or similar coverage area without undue interference.

4.3 The 3G Standard

The 3G standard is being created against the general background of user expectations and industry imperatives. In theory, a single standard could perhaps be created with a single standards body, including all interested parties. In practice, there are different interests and starting points, in particular the desire to evolve from various second generation start points. Figure 4.1 shows the estimated market share (customer numbers, world-wide) by technology type. While the GSM technology dominates this chart, there are some important other starting points for evolution to 3G, including the TDMA (DAMPS) system, used widely in North America, PDC used in Japan and the CDMAone standard used in North America and the Far East. To take account of these interests, within the ITU the concept of a family of standards, known as IMT2000, was developed.

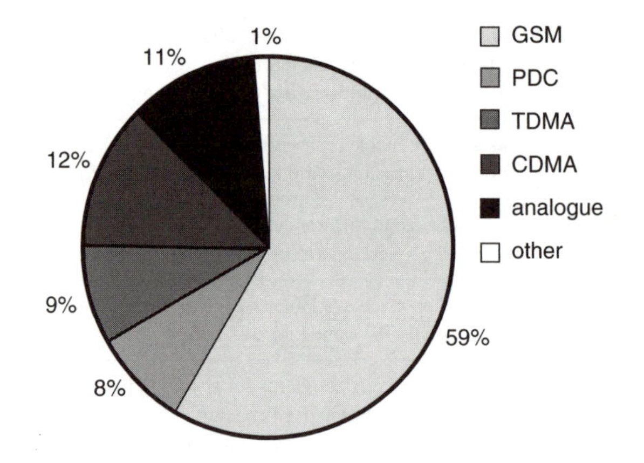

Fig 4.1 Estimate of market share by technology type.

In fact a total of five different air interface standards are included in this IMT2000 family, these being:

- CDMA-DS (direct sequence) — specified by 3GPP, also known as UTRA, FDD mode;
- CDMA-MC (multi-carrier) — specified by 3GPP2, also known as cdma2000;
- CDMA-TDD (time division duplex) — specified by 3GPP, being a combination of the UTRA TDD mode together with the Chinese TD-SCDMA;
- TDMA single carrier — specified by 3GPP, also known as EDGE;
- FDMA/TDMA — specified by ETSI and based on the existing DECT standard.

In addition, there are two different core network standards, one based on evolution from GSM and the other based on evolution from the North American IS41 standard.

While this may seem like a perplexing range of options, there has been some effort to harmonise the different specifications, so that there is some alignment of the key technical parameters within the three CDMA modes, and there are developments to enable each air interface to be used with either of the core network standards. With this set of specifications, each major interest group can find an appropriate evolution path (see Table 4.2).

From an operator perspective, while this approach gives a reasonable upgrade path to 3G, the variety of standards does present some problems. For global roaming, it may be necessary to support several standards, in order to allow access to service in those parts of the world where the home standard has not been implemented. This will require the use of a multi-standard terminal, which currently represents a technology challenge, and is likely to have a cost premium. In addition, it will be necessary to develop cross-standard network interfaces and commercial arrangements. Despite this, the 3G standards landscape is a considerable improvement on the existing 2G multi-standard environment.

Table 4.2 Evolution options for different interest groups.

2G investment	Example operator*	Evolution options
GSM	BTWireless	UTRA FDD and/or UTRA TDD with evolved GSM core. EDGE with GSM core.
DAMPS	ATT Wireless	EDGE with evolved GSM or IS41 core.
CDMAone	Sprint PCS	cdma2000, with IS41 core.
PDC	NTT DoCoMo	UTRA FDD with GSM evolved core.

*Note — the 'example operator' column shows examples of operators with the respective 2G investment and does not imply any declared intent to follow any particular evolution path.

From a supplier perspective, the stabilisation of a standards family provides some good opportunities to enter into markets which have been difficult to access. For example, European suppliers have not been able to penetrate into the Japanese market due to the unique standard in use in Japan, while Japanese suppliers have not capitalised on the opportunity to supply to the world GSM market, due to focus on their home market.

However, the range of standards and standards bodies does present a significant resourcing challenge, for even the biggest suppliers. This is exacerbated by the fact that all standards options are developing in parallel.

For regulators and governments, the openness of a 3G standards set helps in removing artificial trade barriers, and gives an opportunity to attract competitive overseas bidders to enter into the 3G licensing processes.

4.4 3G Standards Developments — Who is Doing What?

4.4.1 3GPP

The GSM community, including ETSI and ANSI-T1 in the USA, along with several Far Eastern standards bodies (ARIB and TTC from Japan, TTA from Korea and CWTS from China), has been brought together by 3GPP [1]. The broad objective of 3GPP is the development of a 3rd generation standard based on a direct sequence wideband CDMA or time division duplex radio interface, coupled into an evolved core network based on GSM Release 98. This makes a significant advance on GSM by providing much higher data rates, but retains the GSM core network, allowing 2G/3G seamless roaming. Several initiatives which were already in place for GSM are also extended significantly, for example, the SIM application tool-kit and the mobile execution environment (MExE), both of which allow a highly customisable terminal environment.

4.4.2 3GPP2

3GPP2 [2] is another global initiative with support from TR.45 in the USA, ARIB and TTC from Japan, and TTA from Korea. It defines the cdma2000 standard which builds on the 2G CDMAone standard. Although data rates are increased roughly in line with the 3GPP standards, the core network is, however, still based on the IS41 core network, making roaming between cdma2000 and 3GPP systems difficult.

4.4.3 ITU

The ITU [3] continues to have a role to play in the development of mobile standards. For some time, the ITU has had a programme of development work under the 'IMT 2000' banner. When the 3GPP and 3GPP2 bodies emerged as strong developers of 3G standards, the ITU agreed to provide an umbrella framework so that these two standards bodies could be recognised by the ITU as developing the detailed specifications, along with ETSI for the enhanced DECT specification. Work inside the ITU and in groups of ITU members has resulted in significant harmonisation efforts which have brought 3GPP and 3GPP2 radio aspects much closer together.

4.4.4 3G.IP

3G.IP was born in May 1999 out of a desire to bring mobile services firmly into the 'Internet age'. Its primary goal was to put the success of 2G mobile systems, particularly those based on the GSM core network, together with the rapid growth of the Internet and the increasing interest in telephony and other real-time services over the Internet. It started modestly, with the objective to make rapid progress by bringing together like-minded companies (3G systems manufacturers and operators) who could develop the core principles for transporting IP-based multimedia services over a 3GPP system based on EDGE or UTRA.

3G.IP was very successful in establishing an agenda towards 'all-IP' and helping 3GPP to develop the first draft architecture, the first step towards a complete IP multimedia system. Once documents with proposals started to find their way into 3GPP, however, it became clear that 3G.IP could not make the desired progress without becoming much more open and allowing other companies to participate. So 3G.IP started to admit other companies towards the end of 1999, becoming completely open early in the year 2000. 3G.IP has no subscription charges and the only prerequisite for joining is the signing of an agreement which may be downloaded from its Web site [4].

4.4.5 IETF

The IETF [5] is focused on the development of protocols which run over an IP stack. As mobile standards migrate increasingly towards IP-based services it becomes more important that the IP service standards take mobile into consideration. There is currently much interest in the IETF in further development of IETF protocols relating to mobile, including call-control protocols such as SIP, mobility services such as mobile IP, and transport efficiency aspects such as IP header compression. The IETF does not consider system aspects and therefore a close dialogue between the mobile standards community and the IETF is required to produce the right standards which help the system to work effectively.

4.4.6 WAP Forum

In 1997 several different manufacturers started to develop proprietary methods for converting Internet content into a more compact format for transmission over expensive and bandwidth-limited mobile data channels. It was during a session in a subgroup of ETSI SMG, responsible for the GSM standards at that time, that some of these manufacturers presented their ideas to the standards community. The ETSI standards community was interested, but disappointed that these ideas were not brought into ETSI, as it was obvious that operators could not support multiple different protocols according to the handsets being used at the time by their subscribers. So these manufacturers were given a clear message that they must work together or their ideas would not take off. The WAP Forum [6] was born later that year, with the objective of defining a standard 'micro-browser' to run on mobile handsets, and a suite of protocols which allow that micro-browser to interact with a network server. WAP micro-browsers can be implemented on a variety of 2G handsets, including GSM, CDMAone and US TDMA handsets. During the year 2000, WAP telephones have started to appear on the market and the WAP Forum is moving on to providing a richer environment for delivering Internet content to mobiles.

4.4.7 Mobile Wireless Internet Forum (MWIF)

While 3G.IP focused on the delivery of an IP multimedia system which builds on 3GPP Release 99, MWIF [7] started early in the year 2000 to study how the existing Internet paradigm could be made mobile. MWIF members come from 3G operators and manufacturers, including mainstream Internet systems suppliers. MWIF also includes the ISP community. MWIF is now producing a proposal for a functional architecture which will be fed into 3GPP and 3GPP2 later this year. The MWIF approach is to define a solution which is independent of access technology and is, in general, much more radical than has been discussed in 3GPP to date, hinting at changing as many as possible of the protocols from today's telecommunications protocols to IP-based protocols.

4.5 The Move to IP Architecture and Standards Impact

Over the past couple of years there has been increasing interest in moving to generalised IP-based architectures and protocols and avoiding expensive specialist protocols and architectures for different implementations of the same service for mobile, fixed, tetherless, etc. In the 3G standards world, one of the most significant impacts is the use of 'voice over IP' technology within the mobile networks. 3G.IP developed an initial architecture which put together the IP call server approach used by both H.323 and SIP with the mobile architecture developed for 3GPP. Instead of carrying voice traffic through expensive MSCs, IP links are established to call servers using the basic IP bearer capability of the mobile network. The call server can then perform largely the same job as in the 'fixed' ISP-based networks, instructing a gateway to set up a call using 'voice over IP' or over the PSTN, and then instructing the mobile to establish another IP bearer channel with higher bandwidth to the gateway for transport of voice traffic.

For 3GPP, however, it is not as simple a process as this. There are a number of aspects which need detailed consideration inside the 3GPP standards process:

- the security and authentication mechanisms have to be examined to ensure that fraud and eavesdropping are avoided, while maintaining the ability to invoke lawful interception;

- the IP bearer channels for both signalling and voice traffic need to be of high enough quality to ensure that calls proceed without failure and that voice quality is equivalent to the circuit-switched world;
- the service or feature interaction between the call control model and additional services provided by operators, such as prepay, VPN, number translation, etc, needs to be handled to offer the same look and feel as the circuit-switched world;
- when roaming, all interactions must be handled smoothly so that the user does not perceive any difference between roaming on circuit-switched networks and roaming on IP-based networks.

The overall architecture for 3GPP has now been defined, and the additional functions required in the 3GPP core network have been termed the 'IP multimedia sub-system'.

Having addressed these considerations, there are the specific protocols to be considered. 3G.IP made an early decision to press for the use of the IETF session initiation protocol (SIP) as a standardised protocol to be used between the user equipment and the call server. This decision was also adopted by 3GPP. In principle, this means that the IETF standard SIP can just be slotted into place, but, in practice, there will need to be changes made to SIP to accommodate the needs of a mobile network. 3GPP has decided that it will encourage the development of these extensions within the IETF rather than making a 3GPP-specific version of SIP. Since the IETF will concern itself only with the protocol aspects, it is necessary for 3GPP to develop the overall system framework and develop the necessary additional function descriptions before the IETF can work out the details of the required extensions.

Overall, this means that the mobile standards landscape is moving from a position where all mobile standards for a particular system are developed in one standards committee structure to one where different standards structures must co-operate to develop the complete solution. The IP architecture is a particular challenge because the companies involved have to bring together small low-power mobile devices having hard-coded implementations with fast-moving IP standards which expect the latest code to be downloadable on to the end-user device.

4.6 Conclusions and Future Outlook on Mobile Standards

There can be little doubt that the pace of change in mobile is accelerating. Can the standards process keep up and continue to deliver the benefits of standardisation without delaying the introduction of new products and services? The future success of standardisation will be dependent on the ability to position standards in order to protect investments while giving freedom to innovate with new developments.

There is an increased speed of change and an expectation of ever shorter development lead times. However, this needs to be counterbalanced by the very high investment costs in introducing new network technologies, e.g. in the UK alone, the 3G licence cost to BTWireless of £4bn, plus an expected network investment of a further £2bn, mitigates against a high rate of change. Putting together the world-wide costs of 3G and taking into account all of the different competing operators across all regions will result in a cost running to many hundreds of billions of pounds.

The key is to provide basic standards which provide a sound and stable platform, protecting the fundamental investments, while enabling a change of functionality and capabilities quickly and at low incremental costs. Some of the current developments will help us move to this paradigm:

- creation of linked standards families, giving evolution options and choices for implementation, e.g. the reuse of the GPRS core network in the 3GPP standards suite enables service and investment continuity during the transition to 3G — for this to work well, a long-term vision is required, such that the basic architecture decisions support future evolution;
- the move to IP-based architectures, where services applications can be developed quickly and supported on a simple, universal protocol structure — the creation of application programming interfaces (APIs) at both the network and the terminal is also a key enabler for this;
- the emergence of techniques such as software radio, where terminals and infrastructure can be reconfigured by downloading or software upgrade to a new standard.

Another factor relating to the rate of change is the way in which standards are developed. This is now a resource-intensive (but vital) part of the development process. Already, standards are created in cyberspace, with electronic meetings, e-document development and e-voting processes. Even so, the lead time for significant standards developments, such as CAMEL, GPRS or 3G is still of the order of 3-5 years. Even relatively small enhancements can take a year to standardise and a further year before deployment. So, one challenge for the industry is to streamline and automate the standards process, with increased use of development tools.

Standards can no longer be developed by small isolated committees. To produce high-quality mobile standards resulting in high-quality global mobile systems with full IP multimedia capability requires development within a number of different global standards organisations. The management and co-ordination of input into all the various forums is a big challenge for all companies in the industry, and, in particular, ensuring that deals struck between companies in one forum are upheld in a different forum will be very testing for all concerned. It is only the major standards players who will be in a position to drive results across all communities and create the right global mobile standards to create and enhance value for their shareholders.

Can standards keep up, and be created fast enough to meet commercial needs? Well, it is the commercial needs which drive the speed of standardisation, rather than vice versa. The difference perhaps is that standards are driven by the overall industry commercial needs, often being a compromise, and perhaps delaying those who would benefit from a very high standards churn rate.

Overall, the outlook for mobile standards is highly positive — they are an intrinsic and essential component of network and terminal developments, and can clearly be seen to have played a major part in the success and growth of the mobile industry world-wide.

References

1 3GPP Web site — http://www.3gpp.org

2 3GPP2 Web site — http://www.3gpp2.org

3 ITU Web site — http://www.itu.int/

4 3G.IP Web site — http://www.3gip.org

5 IETF Web site — http://www.ietf.org

6 WAP Forum Web site — http://www.wapforum.org

7 MWIF Web site — http://www.mwif.org

5

GSM TO UMTS — ARCHITECTURE EVOLUTION TO SUPPORT MULTIMEDIA

N C Lobley

5.1 Introduction

The growth of mobile communications has been phenomenal with mobile penetration of predominantly voice terminals reaching significant levels within many countries. The Internet and the growth in data traffic within networks has also created additional growth within wired telecommunications networks. As networks aim towards their ultimate capacity, it is clear that the future traffic growth area lies with data applications. Although voice will remain the main application, users will, nevertheless, expect significant additional information delivery capability from their terminals irrespective of whether they are fixed or mobile. This chapter explains how the next generation of mobile (UMTS) can be launched and then carries on to propose how the networks can be evolved to deliver the voice and data applications based upon IP technologies.

5.1.1 The Starting Point

The allocation in the early 1990s of spectrum for third generation mobile really spurred on the technology solutions to deliver higher bit rate communications in the mobile arena which had been restricted previously to low-rate voice and simple (<10 kbit/s) circuit-switched data. A variety of drivers and enabling technologies were tabled to provide the high-bandwidth network capabilities to support the greater bandwidths expected from the advanced third generation radio technologies. The proposals included the use of B-ISDN based techniques with ATM, use of pure N-ISDN technology from both wired and wireless networks, along with enhancements of the GSM architecture.

5.1.2 The Conceptual Architecture

The issue of incorporating new radio technology has been a significant one for operators who have had to migrate from first generation (analogue) systems to second generation (digital) systems such as GSM. The development of the W-CDMA and TDD radio technologies and their associated radio access networks (RANs) would take a significant amount of time and effort. This factor, combined with proposals to develop completely new core networks and architectures would have severely delayed the availability of third generation communications and consequently a pragmatic approach to the core network architecture was taken. The ensuing architecture grafted the new UTRAN aspects on to the 'front end' of an 'evolved' GSM/GPRS phase 2+ core network, comprising mobile switching centres (MSCs) and GPRS support nodes (GSNs). This resulted in the concept architecture of Fig 5.1. This solution enables operators who have GSM networks, and also

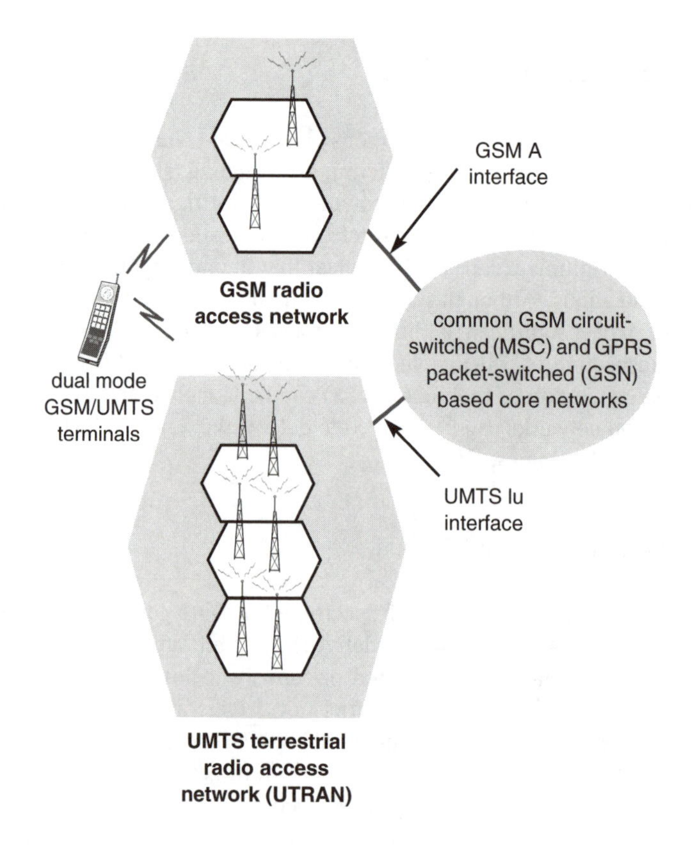

Fig 5.1 The conceptual architecture for UMTS.

suppliers who have core network product lines offering GSM capability, to minimise the technical changes from the contemporary GSM infrastructure. This approach also reapplies the tried-and-tested GSM roaming, charging, signalling and service mechanisms to UMTS.

5.2 The UMTS Launch Architecture

The detailed technical developments added to the GSM core network architecture were predominantly to support the UTRAN aspects which, within the 1999 Release of 3GPP standards (3GPP R99, also known as 3GPP R3) [1], support the W-CDMA radio aspects [2].

5.2.1 The UTRAN Elements

A completely new access network architecture was developed for the revolutionary radio access mechanisms, which took on board the high-speed switching capabilities of ATM, the evolvable support for both W-CDMA and TDMA, as well as delivering standard open interfaces within the radio network. Figure 5.2 gives an overview of the UTRAN architecture.

The basic functional blocks of the UTRAN architecture are the node B and the radio network controller (RNC). The node B can be said to be roughly equivalent to the GSM BTS in linking the antenna site to the network. The node B functions include the radio and modulation/spreading aspects along with channel coding (forward error correction) and some of the combining/splitting functions for soft

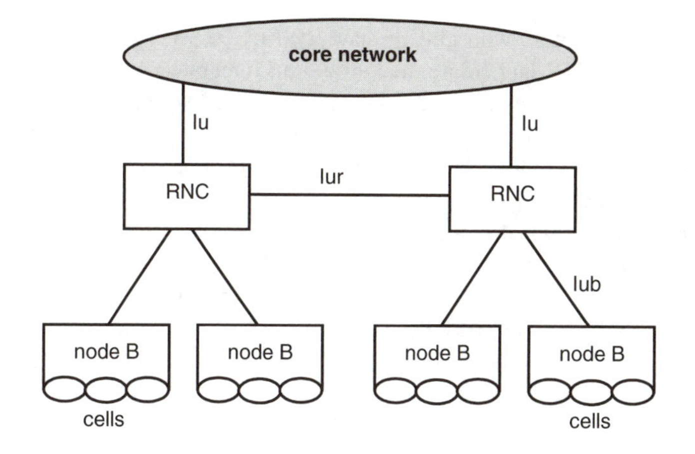

Fig 5.2 The UTRAN architecture for UMTS.

handover. It should be noted that soft handover and micro-diversity is where multiple, differently located receivers are added/removed and used to provide gain in the received signal. The principal requirement with the node B is to minimise the cost/functionality as the network could comprise a large number of node B 'nodes'. The naming of the node B is a result of compromise in the standards process.

The RNC is roughly equivalent at a peer level to the GSM BSC. It is responsible for controlling the resources associated with a number of nodes B, and for negotiating with the core network for aspects such as bearers and quality of service (QoS).

The RNCs and nodes B are connected to each other and to the core network with three interfaces as shown in Fig 5.2. The Iu interface connects each RNC to the core network (similar to the GSM BSS A and Gb interfaces connecting the BSS to the core network) and is responsible for control of handover/reselection, bearer control and negotiation. The Iub connects each node B to its controlling RNC (a similar concept to the Abis interface in GSM) and enables distribution of the radio network functionality. The Iur provides a logically meshed network of RNCs (no GSM equivalent) and is predominantly used for mobility-specific radio reasons which relate to the soft handover and relocation processes.

The interfaces developed within and between the UTRAN and the evolved GSM/GPRS core network use the ATM lower layers combined with common higher level control protocols for both the circuit-switched (CS) and packet-switched (PS) domains [3, 4]. The common RANAP protocol provides resource control and allocation, handover and bearer requirements, control, and certain charging and control linkages between the UTRAN and core network.

5.2.2 Network Launch Configurations

Those network operators who already have GSM/GPRS networks have two basic choices for the UMTS launch architecture — an integrated solution or an overlay solution (as shown in Figs 5.3 and 5.4). The integrated solution (Fig 5.3) sees the current GSM/GPRS core network aspects upgraded and reused with the same switching (MSC) and routing (GSN) elements used for both GSM and UMTS radio. The new UTRAN is connected to this network using the Iu interface. This approach allows the reapplication of common O&M systems, service-delivery mechanisms, switch sites, and platforms; however, the capacity, performance and network growth impacts of connecting relatively new and unproven W-CDMA access technology to a live, service-providing network need careful assessment.

The overlay solution relies upon operators using a different (overlay) network of switching (MSC) and routing (GSN) elements to support the UMTS radio. The overlay solution enables a parallel independent development of the UMTS access with lower risks to the live GSM/GPRS network. The 3G MSCs need similar service delivery mechanisms to be developed (to enable users to receive equivalent

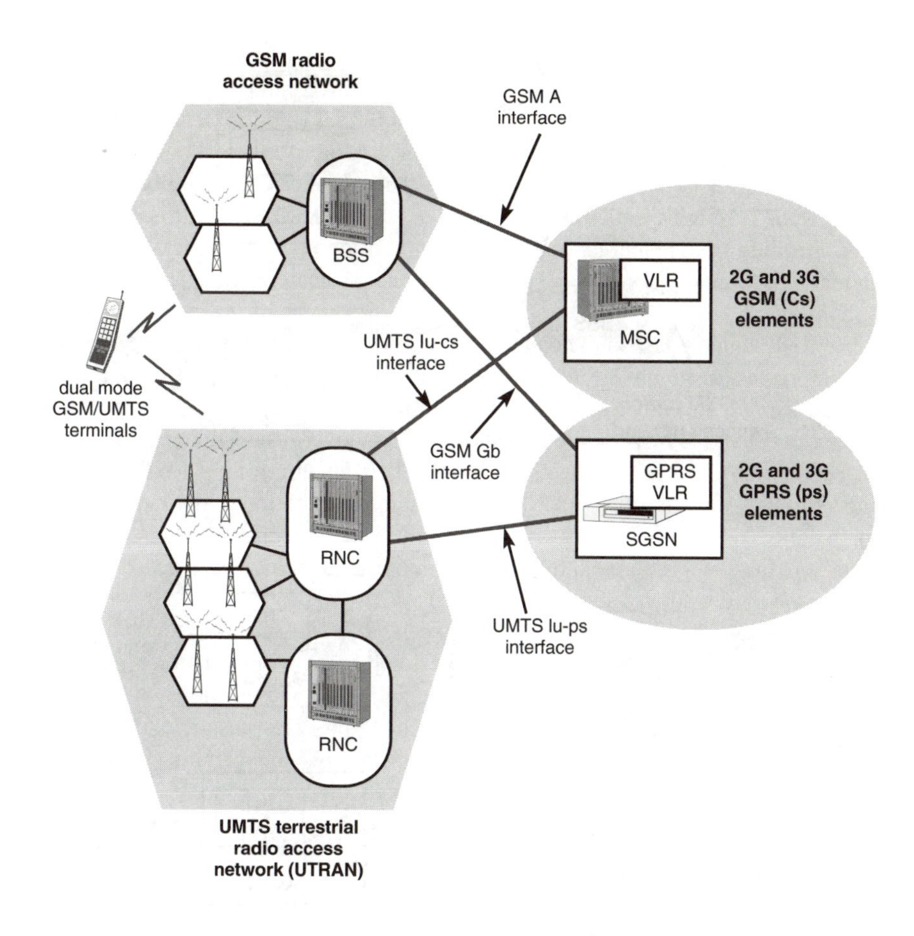

Fig 5.3 Integrated option for UMTS launch.

services via 3G as well as 2G), as well as O&M and site/platform capacity to support the new infrastructure. The benefit of the overlay is that operators can roll out and develop the UMTS aspects of the network in relative isolation from the live, revenue-earning 2G network. The open Iu (UTRAN core networks) interface enables different core or access network suppliers to be chosen and so enable operators to try an alternate supplier for 3G, or to source a single-supplier, turnkey network.

At the time of writing Finnish, German, Dutch and UK operators are actively procuring, developing and rolling out UMTS architectures based upon the evolved GSM approach.

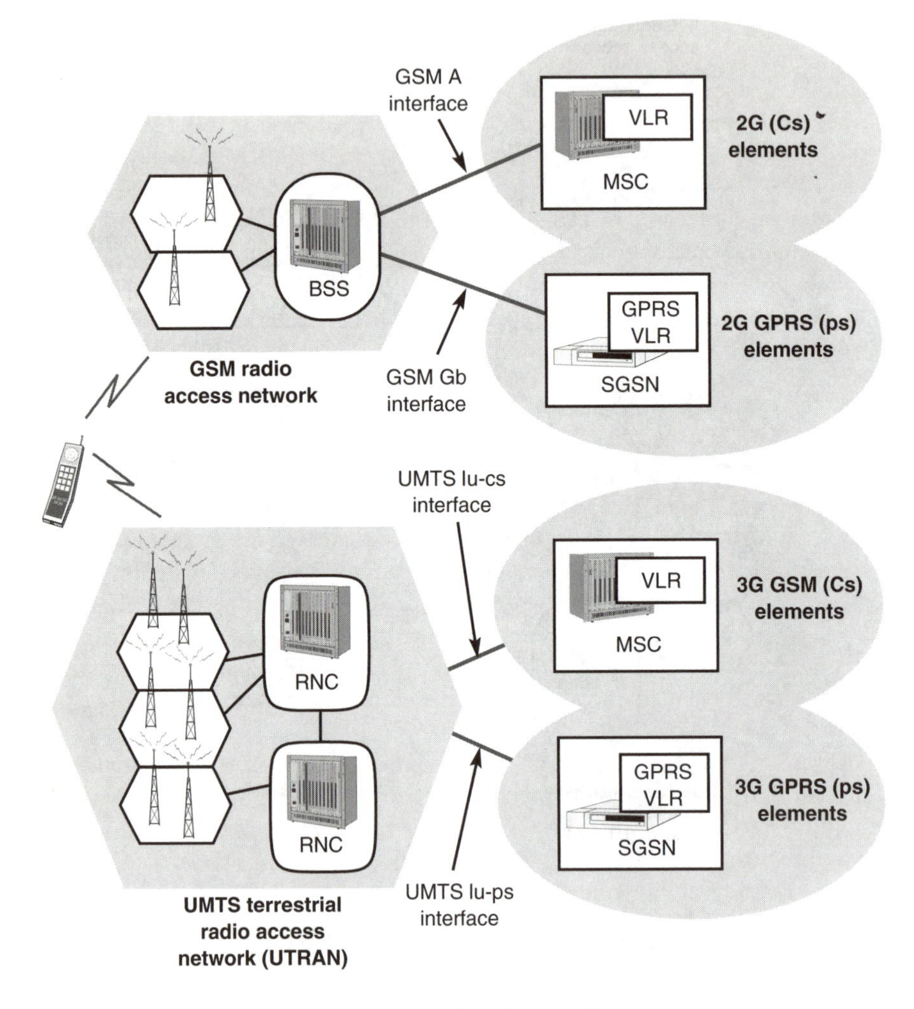

Fig 5.4 Overlay options for UMTS launch.

5.3 Network Requirements and Influencing Technologies

As operators start commercial service of UMTS based upon the launch architecture, they are also looking at the onward evolution paths and opportunities open to them. Recent years have seen extensive changes in the fields of telecommunications. The WWW and Internet protocols have made a significant impact upon the world of user services, data availability and network connectivity. The growth in usage of the Internet via dial-up PSTN access has been phenomenal and the higher speed access available, from technologies such as ADSL, should boost wired connectivity and further expand the Internet 'bubble'. In parallel the mobile world has grown from a niche, high-cost, low-volume market to a high-volume mass-market product with significant numbers of users, ranging from simple pre-pay through to complex data and voice corporate users. Initial moves towards the mobile data wave are indicated by the growth in SMS usage, significant sales of WAP phones [5] and the launch of operators' GPRS networks offering true IP access from mobile. The higher bandwidth capabilities of UMTS should fuel the growth in both data applications and the simple 'voice-only applications' which have predominantly been the bread and butter of mobile operators to date.

In parallel to pure data applications and voice, operators of launch UMTS networks are likely to investigate communications mechanisms such as video and multimedia. These have generally been slow to enter the mobile and fixed markets due to legacy network transmission systems, cellular user bandwidth and terminal availability.

These limiting factors are starting to be removed by packet-based (Internet) networks, UMTS high-speed cellular bandwidth, and the wider availability of feature-rich mobile terminals. Within the mobile market, all the terminals conform to standards to enable cellular operation and generally have a short lifetime in the market before replacement (as consumer items they typically are replaced every eighteen months to two years). The relatively short span and consumer nature of the mobile terminals (combined with cost/volume benefits) enable new standard features to reach the market-place in relatively short time to access features via the network.

As well as new service-capability opportunities, operators are also looking to reduce the cost of operation and ownership of their networks, including increased flexibility in the transmission of information within their networks. They are also constantly seeking greater differentiation from their competitors. While GSM was a great success in its ability to provide standard roaming — including the support for standard service features — the opportunity to create operator-specific features was to a great extent limited by the close linkage of speech services to the complex mobile-specific switching platforms (MSC). Capabilities such as CAMEL [6] and the SIM tool-kit have eased the service issue to a certain extent; however, the UMTS focus on delivering data and IP capabilities suggests more operator-specific services will become available. The services can be delivered by re-applying contemporary IP-based multimedia techniques such as those developed in the IETF [7].

5.4 Post-Launch — the Evolution Drivers

As operators progress the launch of their UMTS networks based upon the evolved GSM/GPRS approach (based on 3GPP R99/R3 standards), they are looking at the next steps and how to move forward. With the trial and introduction of W-CDMA into the network completed, the next stage is to build service capability and cost optimisations into the core network.

From the fixed telecommunications circuit-switched network perspective, a number of developments are ongoing to give operator's greater flexibility in the deployment of networks. Distributed processing has enabled the separation of pure switch/routing functionality away from the control mechanisms [8]. The separation of contemporary switch mechanisms into media gateways (MGWs) (containing switching, transcoding and user-plane transmission aspects) and media gateway control functions (MGCFs) (containing switch and service control functionality), connected via standard interfaces (e.g. H.248/Megaco), will enable operators to increase the service delivery and control parts of their networks in relative isolation to the growth of the user traffic parts of the network. Figure 5.5 illustrates the concept behind the distributed processing and switching mechanisms offered by H.248 and Megaco. This separation additionally gives operators the ability to increase the control and bearer aspects of their network in relative isolation and also enables procurement towards distributed networks with controller and gateway procured from separate suppliers, enabling a real progression towards call server 'farms' connected to 'pools' of resource control and switching.

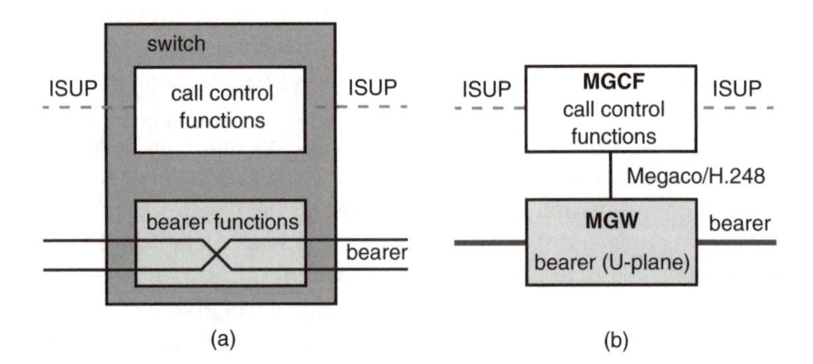

Fig 5.5 Conventional switch (a), and decomposed switch using the H.248/Megaco concept (b).

While contemporary telephony networks are relatively low-cost service-specific networks delivering mass-market voice, there are several initiatives ongoing to provide speech and multimedia services over IP-based networks. Standards initiatives, such as TIPHON [9] and SIP [7], have moved forward the ability of operators to deliver services including voice and multimedia which are neither limited nor restricted to the old fixed bandwidth and functionality limitations of 64 kbit/s TDM technology.

5.4.1 Network Requirements and Influencing Technologies in the Evolving UMTS Network

The influence of the service and cost drivers, and the technological opportunities from the wider telecommunications and IP areas, has been taken into account with the specification of the post-launch standards for UMTS.

The circuit-switched arena has adopted the distributed concepts of H.248/ Megaco while the packet-switched domain has been enhanced to adopt VoIP/ multimedia-over-IP concepts.

Within the traditional circuit-switched part of the network the main area of focus has been the separation of the MSC into the separate components of MSC server and MGW (see Fig 5.6). The MSC server comprises all the mobile-specific call, service and mobility controls, and uses H.248 operations to control the MGW. The MSC server uses H.248 to interact with the MGW for circuit-switched communications, including enhancements to provide the allocation of switching and transcoding functionality for mobile speech and circuit-switched data communications, the support of path changes for handover and SRNS relocation (path optimisation within the UTRAN), and the addition and deletion of multiple paths for circuit-switched communications.

As illustrated in Fig 5.6 this development enables multiple interactions between the MSC servers and the different MGWs within the network.

This enables flexible transcoder aspects such as tandem-free operation (TFO) and transcoder-free operation (TrFO) to be supported, thus reducing the transmission costs to the network operator by allowing lower rate (mobile coded rate) speech to be carried across the network. For example, within Fig 5.6 MGW3 could be the gateway interconnecting with the fixed network while MGW1 links to the UTRAN. In such a scenario the speech transcoder for a voice call would reside in MGW3, thus enabling adaptive multi-rate (AMR) coded speech to be carried through the mobile network at low rate until MGW3 transcodes AMR back to 64 kbit/s coding.

From the packet-switched domain perspective, the key development is the support of real-time communications over evolved GPRS. The packet components of the launch architecture should support a variety of quality-of-service (QoS) mechanisms which will allow best-effort, conversational streaming and interactive IP-based communications.

Beyond launch, these introductory features will be enhanced to deliver continuity of service (such as a change of SGSN during an active QoS-based session), taking into account those mobility aspects that must be fully supported. Linkage of the GPRS elements to the CAMEL features will enable services such as GPRS pre-pay to be delivered.

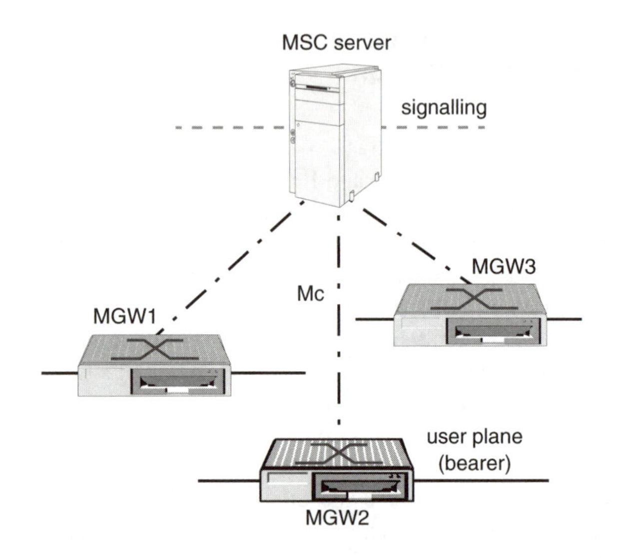

Fig 5.6 The MSC server concept in UMTS.

With a true industry-grade cellular QoS mechanism in place the challenge will be the addition of voice and multimedia-over-IP mechanisms via evolved GPRS. This capability will enable operators to converge their networks from the traditional parallel packet- and circuit-switched infrastructure towards an 'all-IP' operation. The addition of QoS to the packet-switched domain enables the addition of call servers and gateways to enable mobile VoIP users to access legacy telephony with 'operator-quality' voice. Within the 3GPP groups, service developments for VoIP and multimedia over IP have been termed 'IM services' with the associated network elements to support this known as the IM CN (core network) sub-system. These network elements include call servers (known as call-state control functions) and media gateways. A conceptual view of the IM core network sub-system is shown in Fig 5.7.

It should be noted that similar developments are also taking place within some contemporary fixed IP networks, which just replace the GPRS packet elements, radio access and mobile terminal with higher bandwidth access (e.g ADSL) and wired IP terminals respectively.

The parallel development of the IM core network sub-system and wired IP voice networks should give operators and suppliers benefit in providing access-independent equipment procurement and provision, thus reducing cost and network/equipment management and development overheads.

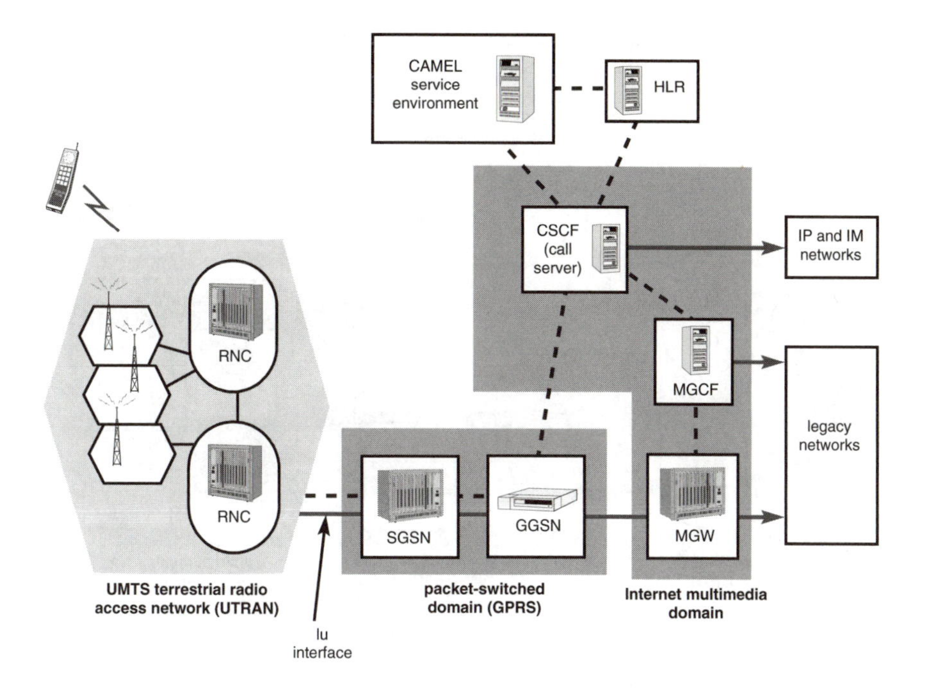

Fig 5.7 Concept of VoIP MGW and call server, via enhanced GPRS.

The call-state control function (CSCF) acts as a call server and handles call signalling, while the MGCF and MGW provide legacy interconnection. Connection with other IM-enabled networks can be handled via the CSCF interconnections.

5.4.2 The IP Multimedia Sub-System Architecture

To cater for the mobile specific nature of the users who expect to be able to roam between and within networks to access their services, the basic architecture of Fig 5.7 has been enhanced further as shown in Fig 5.8.

The GSM/GPRS HLR has been evolved with the inclusion of additional user data to cater for the IM aspects and has been renamed the 'home subscriber server'. The HSS also drives towards the 'access-independent' nature of the network to cater for users who are served by IP networks other than GPRS. Access independence enables operators to utilise the IM aspects of the network across a variety of accesses, not just UMTS. Both radio and wired (e.g. xDSL, packet-cable) access could be used. The CSCF has been functionally decomposed into three main areas:

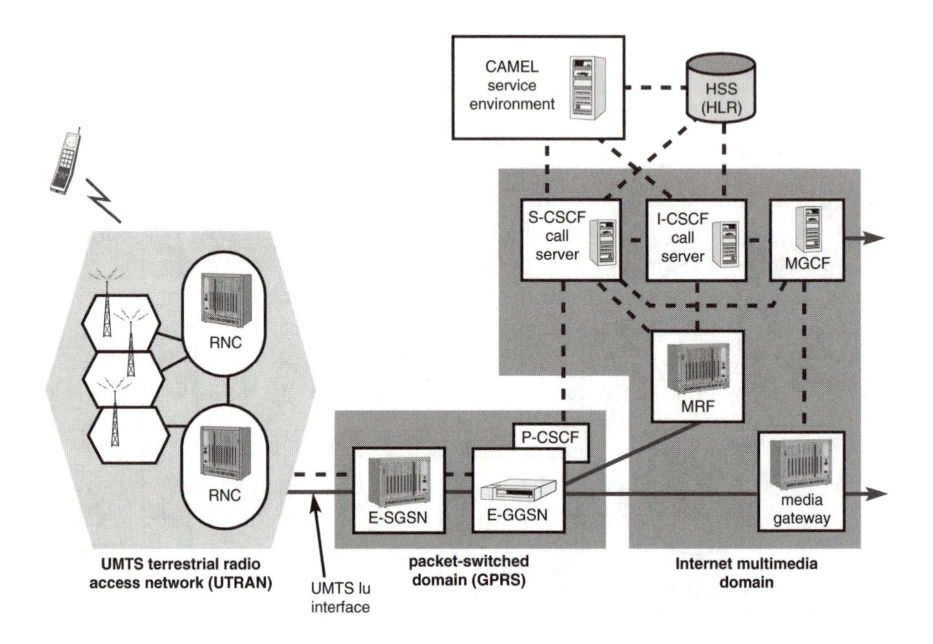

Fig 5.8 A detailed view of the IM core network sub-system.

- the serving CSCF (S-CSCF) which holds user-specific data such as details of some of the services supported (downloaded from the HSS) — the S-CSCF also terminates the call/session signalling from the mobile terminal (Ue) and interacts with the applications and services area;
- the interrogating CSCF (I-CSCF) which interrogates the HSS during mobile terminated communications set-up, to determine the S-CSCF catering for the mobile;
- the proxy CSCF (P-CSCF) which performs bridging of the signalling (acting in a similar way to that of a firewall) between the terminal (Ue) and the S-CSCF — the proxy is also involved with bearer confirmation and decision, and resource aspects during session bearer establishment.

The MGW acts in a similar way to the circuit-switched area, providing the transcoders and interconnection with legacy networks. For the IM aspects, the MGW does not need to provide the complex handover and SRNS relocation aspects, used with the MSC server as all core network mobility-related aspects are provided by the GPRS network elements (SGSN/GGSN). The media resource function (MRF) can be used to provide IM-specific functions such as multiparty call and media-conferencing functions, including relations with the bearer and service validation.

The MGCF is applied in the same way as for wired VoIP networks in providing IP-specific-call/session-signalling-to-legacy-network-signalling conversion (e.g. SIP to SUP).

A variety of options were available for the delivery mechanisms of the IM core network sub-system in terms of terminal (user equipment (Ue)) to call server (CSCF) signalling, links to the applications space, and also IP version used. For the Ue-to-CSCF signalling, the two main contenders were SIP signalling [7] and ITU H.323 multimedia signalling. The decision was taken to use SIP due both to the benefits of closer linkage with the IETF, with its flexibility for voice and multimedia applications, and the opportunity to influence development to allow mobile-specific developments to be taken into account. IP version 6 (IPv6) was chosen for use within the IM core network sub-system to allow the use of v6 from day one, thus minimising IPv4-to-IPv6 upgrade and compatibility issues — this also promotes the early introduction of the wider advantages of IPv6 into the industry.

A number of options and decisions still need to be made regarding the core network signalling relationships. Between the HSS and the CSCF, the main contenders are LDAP and evolved MAP signalling using IP — the trade-off between the database/data transfer protocols and wide industry availability of LDAP and the mobile-specific, evolvable and optimised MAP aspects are still being assessed within the 3GPP organisation. The architecture enables the serving CSCF to deliver services and features to users as a stand-alone entity or networked to the applications and services entities. At the time of writing the possible interfaces between the CSCF and the service platforms are as follows.

- CAP

  This interface provides CAMEL-based services offered on the CAMEL CSE platforms. They are invoked by a service switching function (SSF) and supported by the CAP protocol [6]. A 'softSSF' in (or on top of) the CSCF is required for mapping of the SIP state machine in CSCF to the CAMEL BCSM (basic call state machine). This interface eases the re-application of circuit-switch domain services delivered by CAMEL using the IM components as well as roaming with 2G circuit switched.

- SIP

  This interface provides services offered by SIP application servers and SIP-based multimedia service platforms. These services are directly invoked by the CSCF as a SIP server itself. The use of SIP as the interface enables the serving CSCF to interact with other (service-specific) CSCFs, to deliver capability.

- OSA (open service access)

  This interface proposes to provide applications that are independent from the underlying network technology, particularly those supplied by 3rd party application developers or service providers (for more details about OSA, see Chapter 8).

5.4.3 User and Service Delivery Considerations of IM

As UMTS operators increase their subscriber base using the Release 1999 solution, careful consideration needs to be given to the user and service migration towards the IM capabilities. While legacy terminals and simple voice communications will remain supported by the MSC technology for many years to come, a movement of subscriber numbers and services to IM will occur and will need to be carefully managed.

Many parties have proposed that the IM area should be offered as a complete stand-alone, new capability sub-system, but the commercial reality of offering user services needs to be taken into account. Users already have a mobile telephone number which for IM they could probably supplement with Internet-type addresses and identities; however, especially at launch of IM, the bulk of traffic will be to and from the legacy PSTN networks. Users will not be prepared to be given a new 'IM-specific' number and identity when they already have one from 2G cellular. Users will wish to receive and make communications not only when in the coverage of the IM-enabled UMTS radio, but also when outside this area, for example when covered by GSM or pre-IM UMTS coverage.

The use of SIP signalling between the terminal and network/service delivery components opens the door for operators to move away from the constraints of legacy ISDN-based systems and offers users contemporary services, such as voice, but supplemented with additional value-added services. This will be possible due to the ability to pass additional information via IP signalling to the mobile terminal from the network or application. Figure 5.9 gives an example of the different networks from which users will wish to receive their services. As well as the real-time services such as voice, the users can still receive basic GPRS IP connectivity via both GSM and UMTS radio — however, this is not explicitly shown in Fig 5.9. The four key points to note for the service delivery mechanisms required for the different users are given below.

- User in IM-enabled home UMTS network coverage (Fig 5.9, Area 1)

  The user is served by the home network CSCF using home network UMTS coverage. The user should be able to access a whole variety of capabilities including those evolved from the old legacy services. For example, a key component could be voice as part of a multimedia session. When the called party can communicate via voice and video, the IM aspects shall support this; however, if the called party can only support voice communications (e.g. called party is PSTN or IM user with restricted bandwidth), the IM shall just deliver these capabilities. It is worth noting that, when IM aspects initially are available, a large proportion of communications will be to and from legacy networks, and IM mechanisms will have to support certain services that users used to receive via the MSC.

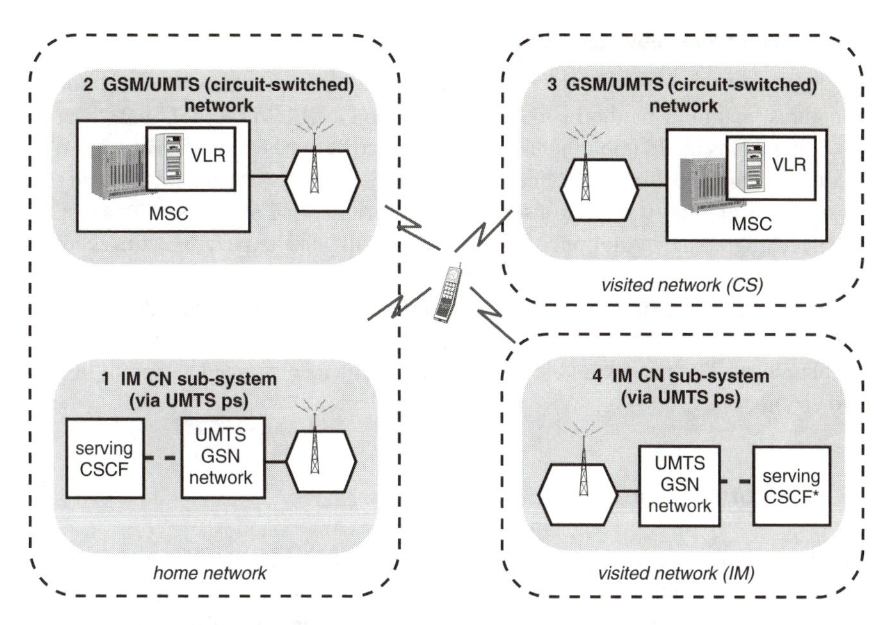

Fig 5.9 Service delivery to IM users.

- User not covered by IM-enabled UMTS coverage but covered by home GSM (or UMTS MSC) (Fig 5.9, Area 2)

  As operators continue to roll out networks, it is obvious that providing the depth of coverage equal to mass-market population coverage achieved with GSM 900/1800 will not be possible for many years. With this in mind the early releases of UMTS enabled circuit-switched users to access similar services via GSM or UMTS (e.g. telephony). It is not feasible for IM to be delivered over the GSM 900/1800 networks so operators need to deliver solutions that enable users to access some of their services via the CS aspects (MSC). Area 2 of Fig 5.9 shows an IM user who is out of coverage of IM-enabled radio but able to access certain features via the GSM MSC. For example IP-based video may not be possible, while telephony could be provided as an alternative to VoIP-type telephony while still under the coverage of the home operator. The user would still be reachable by their same identity (MSISDN) and also be able to modify their same services while out of 'IM' coverage.

- User in IM-enabled UMTS coverage while roaming (Fig 5.9, Area 4)

  When users roam to other IM-enabled networks they should be able to access their home services. The local CSCF shown in the visited network can be used to deliver services offered by the visited network direct to the user. For the users more personalised services the home network based CSCF is used with the SIP signalling passing between the visited network and the home network.

- User roaming to GSM network (Fig 5.9, Area 3)

  When users roam outside the home network to networks that do not support IM features, a similar method is used to the home GSM/UMTS MSC-based service solution. This relies upon the use of MSC-based services and mechanisms. When considering the different service delivery areas for users, the operators need to be careful to gradually migrate users who wish to use IM services. The users will expect a gradual evolution of service and will also expect to gain access to service when outside the IM coverage. This requires very innovative application and development of not just the new IM aspects for UMTS, but also active management of the user experience when covered by older MSC-based technology. The most suitable network mechanisms also need to be put in place to enable this.

5.5 Conclusions

This chapter has shown how operators can move from legacy GSM/GPRS networks towards initial launch of UMTS networks and then on to reach the real 'all-IP' solution of 'IM' and the feature-rich user opportunity that this offers. Operators will see benefits from the cost and capability optimisations in the circuit-switched area as well as the service migration path towards IP-based voice and multimedia applications.

As the race to gain UMTS licences and to roll out networks gains speed, and as operators move from launch phase to consolidation it is obvious that the evolving UMTS architecture will give operators the platform to broaden out the service capability to users. This will not only provide higher speed access to data/Internet services but also create an environment to deliver services which are a long way away from the current simple telephony services we experience and use today. In reality, these new features cannot be created by a 'big bang' approach and some of the limitations of legacy systems must be taken into account when providing the complete package for users. This will ensure that while users can access the feature-rich capabilities offered by IM, they still get usable and meaningful capability when served by legacy networks.

References

1 3GPP Release 1999 UMTS specifications — http://www.3gpp.org

2 3GPP Specification 25.401: '*UTRAN Overall Description*', Release 1999, Version 3.2.0 (1999).

3 3GPP Specification 25.410: '*UTRAN Iu Interface: General Aspects and Principles*', Release 1999, Version 3.2.0 (1999).

4 3GPP Specification 25.413: '*UTRAN Iu Interface: RANAP Signalling*', Release 1999, Version 3.1.0 (1999).

5 WAP Specifications — http://www.wapforum.org

6 3GPP Specification 29.078: '*CAMEL: Stage 3*', Release 1999, Version 3.3.0 (1999).

7 IETF: '*SIP specification*', RFC 2543— http://www.ietf.org/

8 ITU-T: '*Gateway Control Protocol*', H.248 Standard (March 2000).

9 ETSI: '*Telecommunication and Internet Protocol Harmonisation Over Networks (TIPHON): Description of technical issues*', ETS TS 101 300 (1999).

6

VOICE AND INTERNET MULTIMEDIA IN UMTS NETWORKS

M C Bale

6.1 Introduction

The main driver behind 2nd generation digital mobile networks, such as the global system for mobile communications (GSM) [1], was the need to provide a voice telephony service to mobile users. This has been achieved with incredible success. Moreover, GSM has established the starting point from which future mobile networks must evolve and an important benchmark for voice services that the 3rd generation of mobile networks must exceed in terms of functionality and quality.

The Universal Mobile Telecommunications System (UMTS), the 3rd generation network and systems standardised by the 3rd Generation Partnership Programme (3GPP^{TM}) [2], aims to provide voice services that will meet the needs of mobile users. This is being done in collaboration with the International Telecommunications Union (ITU) 'International Mobile Telecommunications — 2000' project [3].

In the initial phase of UMTS, defined by the 3GPP Release 1999 standards, the voice telephony service is essentially an evolution of the GSM voice service that benefits from the 3rd generation technologies adopted for the UMTS Terrestrial Radio Access Network (UTRAN) (see Chapter 11 for more details).

However, the customer's needs for mobile voice telephony must also be considered in the light of the growing demand for mobile Internet multimedia services. In particular, voice will be a feature of many of these multimedia services, e.g. videoconferencing, mobile commerce (mCommerce), games and multimedia messaging. To enable such services, it is important that the voice service is as much part of the mobile Internet as the data and information services with which it will be integrated. This requires a more radical approach to the provision of voice services, one that is more aligned with the Internet and the protocols standardised by the

Internet Engineering Task Force (IETF) [4]. This challenge is being addressed by 3GPP in the production of the Release 4 and 5 standards, and by the IETF in the production of the protocols needed to realise mobile Internet multimedia.

This chapter initially provides an overview of how a voice telephony service is supported by a UMTS network conforming to the 3GPP Release 1999 standards. It then describes how the subsequent 3GPP Release 4 standards allow for the Internet protocol to be used as a bearer service for voice. The chapter then goes on to describe the proposed solution currently being standardised by 3GPP for Internet multimedia services (including voice), known as the Release 5 standards. This solution is illustrated with message sequence flows to show the dynamic aspects of the solution and the application of the various protocols. It is assumed that the reader already has an awareness of GSM and general packet radio service (GPRS) networks.

Work to address the challenges of providing voice and multimedia services in a mobile and wireless Internet environment is progressing rapidly within 3GPP as well as the other bodies producing standards for this area (such as the IETF). However, the reader should be aware that there is still much work to be done, especially at a detailed level. At the time of writing, this chapter reflects current views, which may differ from the actual standards when they are completed. To aid understanding of some of the issues, potential solutions are described, but it should be recognised that these are only illustrative and may not be endorsed as standards in the future.

6.2 Voice in the 3GPP Release 1999 Network

Release 1999 is the first phase of the 3GPP standards for UMTS. This is a completed set of standards that defines a UMTS network able to provide users with voice and data services fully compatible with those of GSM and GPRS. The standards allow users to migrate on to the UMTS and to roam seamlessly between UMTS and GSM/GPRS networks without any loss of capability. It also has the benefit to the network operator of being able to target the introduction of UMTS to specific geographical areas, while relying on existing GSM and GPRS networks to provide coverage in other areas.

Specifically, current GSM networks support voice and low-speed data services that are circuit-switched, so called because the voice or data is carried between users in bearer circuits that are switched into place across the network for a time period, under the control of signalling from the users. In contrast, the GPRS network supports packet-switched data services. For the purposes of this chapter, only the voice services in Release 1999 are described, but the descriptions also apply to low-speed circuit-switched data services.

Figure 6.1 shows the overall network for the support of voice services in the 3GPP Release 1999 standards, and is more fully described in Chapter 5.

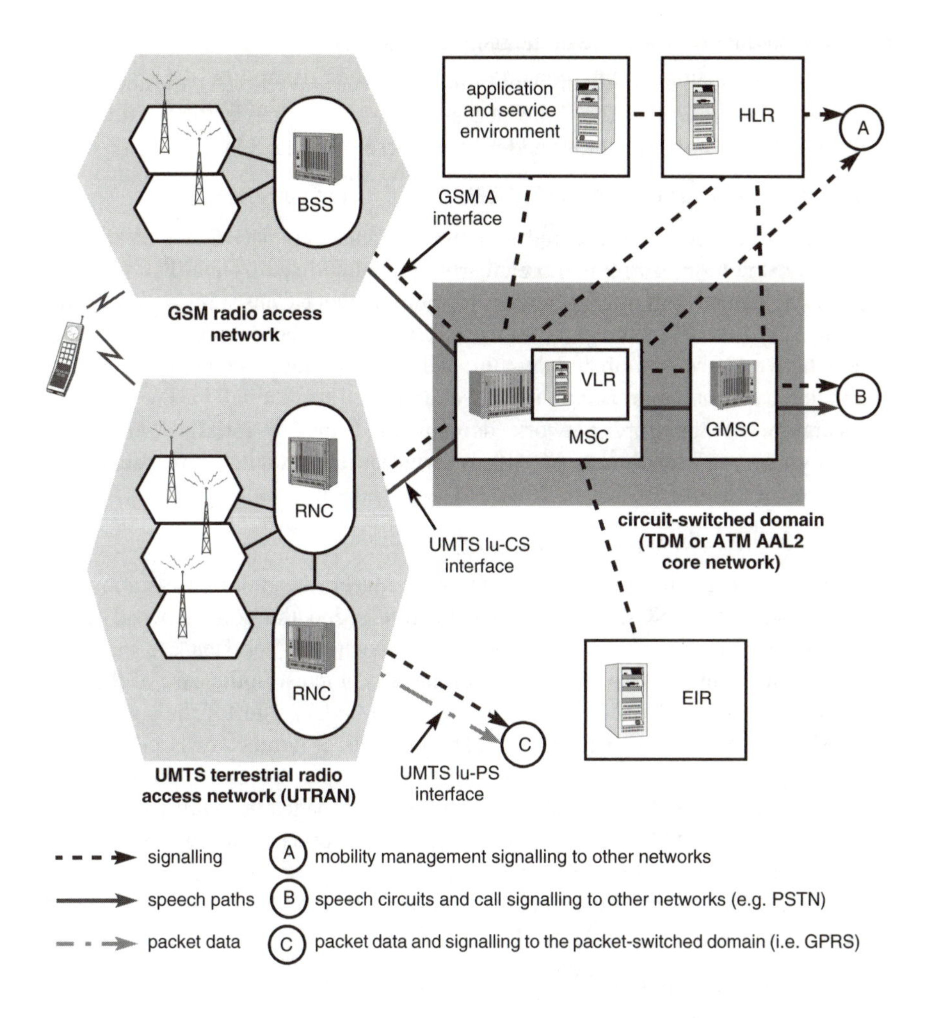

Fig 6.1 3GPP Release 1999 voice network overview.

To achieve compatibility with GSM, the Release 1999 network effectively adopts the GSM core network and service architecture. This has a significant benefit to the network operator since it enables a cost-effective and low-risk evolutionary approach to be taken for the deployment of UMTS. However, in UTRAN, changes in both the architecture and the radio and transport technologies employed result in differences from GSM, but also enable some improvements to the way in which voice services are provided. The main areas where UTRAN affects voice services are described below.

- Improved quality of service in the radio access

  The use of wideband code-division multiple access (WCDMA) and the various modes of operation in the radio access can improve the quality of the voice service in terms of availability and reliability (see Chapter 11).

- Well-defined interface between a UTRAN and a core network

  The interface between the radio network controllers (RNCs) and the core network, the Iu interface, is more clearly defined and open, such that a UTRAN from one vendor will interoperate with a core network from another vendor. The Iu interface itself is separated into the Iu-CS interface between the RNC and the core network circuit-switched domain, and the Iu-PS interface between the RNC and the core network packet-switched domain (not shown in Fig 6.1). The separation of the core network domains and the Iu interface allows the deployment and evolution of voice services independently of packet data services in Release 1999.

- Use of ATM as the transport technology

  ATM is used as the transport technology between the radio base-stations and RNCs, between RNCs, and between the RNCs and the core network (the Iu interface). Both circuit-switched and packet-switched services are carried in ATM cells, using appropriate adaptation layer protocols. In the case of the voice bearer circuits this is ATM adaptation layer 2 (AAL2), and for the signalling is ATM adaptation layer 5 (AAL5). ATM provides a number of benefits in the access network, such as the ability to transport packet- and circuit-switched services with low delay, high bandwidth and manageable quality of service. Conversion of ATM to the circuit-switched time division multiplexed (TDM) technology, if used to switch the voice paths in the core network, can be performed by the mobile switching centre (MSC) or by a gateway function between the RNC and the MSC.

- Speech transcoders located in the core network

  Speech transcoding is performed in the MSC in Release 1999, rather than at the base-station sub-system of the GSM radio access network. The relocation of this function into the core network allows operators to provide lower-cost access-transmission networks, and eases the introduction of transcoder and tandem-free operation.

A significant benefit of retaining the GSM core is that the MSCs can interface to both the UTRAN and existing GSM radio access networks, and more easily support user roaming and in-call handover from the UMTS to GSM networks. Within the core network, the only notable change from GSM is that voice services may be supported either on circuit-switched TDM (as in GSM) or via ATM transport. Again, AAL2 is recommended for providing the voice bearer circuits and switching if ATM is used. Other transport protocols such as ATM adaptation layer 1 or voice-over-IP solutions could in theory be used instead — although these may not meet all the quality-of-service requirements of a Release 1999 network.

In Release 1999, the user's speech is digitally sampled by the mobile user equipment, and then coded for transmission. The default speech coding, which must be supported by all mobile user equipment (terminals) and the UTRAN, is adaptive multi-rate (AMR). The AMR coder supports eight source rates ranging from 4.75 kbit/s^{-1} to 12.2 kbit/s^{-1}, and is rate-controlled which enables it to rapidly switch between these at any point in the call. The AMR coder encodes and decodes the digitally sampled speech to make optimum use of the battery power and bandwidth available, particularly on the radio link between the mobile equipment and the radio base stations (node B). The bit rates are selected depending on the quality of speech required and the quality of the transport provided by the network, and primarily that of the radio link. The AMR coder also supports a low-rate background noise encoding mode to reduce transmission during silence, further reducing bandwidth and battery usage in the user equipment. In addition to AMR, other speech coding may be optionally selected, such as enhanced full rate (EFR) or full rate (FR), as also specified for GSM. Within the core network, the ITU-T Recommendation G.711 on speech coding at 64 kbit/s^{-1} or 56 kbit/s^{-1} is generally used as in the public switched telephony network (PSTN) and GSM core networks. Transcoding from AMR (or other speech coding) to G.711 is performed in the MSC.

If the users' equipment at both ends of a voice call use the same coding, then transcoding to G.711 (or other codings) is not necessary. There are two procedures that can be adopted to remove or reduce transcoding, namely:

- tandem-free operation of transcoders — where inband signalling between the transcoders determines the transcoders in use and allows the transcoders to drop out of the speech circuit if both terminals are using the same speech coding;
- transcoder-free operation — where the mobile terminals negotiate the speech coding during call set-up, and transcoders are only inserted into the speech path if end-to-end compatibility cannot be achieved.

Although considered for Release 1999, it is not until Releases 4 and 5 that standards will be available for tandem-free operation and transcoder-free operation.

As with GSM, signalling from the user to the network broadly falls into two categories — call-related signalling for establishing, maintaining and terminating voice calls, and non-call-related signalling for mobility management (e.g. for location registration, roaming and in-call handover). The signalling protocols and procedures are generally the same as for GSM, although new lower layer protocols provide adaptation to the underlying ATM transport in the UTRAN. Within the core network and for interconnect to other networks, the ITU-T recommendations for Signalling System No 7 (SS7) are used, again with lower-level adaptation layer protocols in the case where ATM transport is used.

Supplementary services, such as call diversion and caller identity, are provided from the MSCs, which also provide tones and announcements to the user. More advanced voice services can be provided from the application and service environment, as described in Chapter 7. A user profile, containing information on the individual's subscribed services, is provisioned into the home location register (HLR) for that user. This is then copied into the corresponding visitor location register (VLR) in the MSC responsible for controlling users' calls, so that their services can be provided as they change location. For billing purposes, call detail records, for example containing information on call duration and destination, are generated by the MSCs and sent to the operator's billing engine. Information may also be collected from the HLR for billing purposes. The MSC also communicates with an equipment identification register (EIR), for example to validate whether the mobile terminal is a stolen one.

With the Release 1999 standards completed, it is anticipated that UMTS Release 1999 voice networks will be operational by 2002, with operators already beginning to deploy network and UTRAN equipment in order to meet this date. However, it is not until the second phase of UMTS standards that support for other real-time multimedia services is defined.

6.3 Migration to Internet Protocol Voice Bearers — 3GPP Release 4

The following two phases of UMTS evolution specify how voice and multimedia can be supported by an Internet protocol (IP) transport service. Currently, two phases are defined:

- Release 4, which includes the migration of the Release 1999 circuit-switched domain core network and services to an IP transport;
- Release 5, which takes a more radical approach to the introduction of conversational and interactive multimedia services on to an end-to-end IP transport provided by an enhanced general packet radio service in the packet-switched domain — this is described further in section 6.4.

These releases were previously known singly as Release 2000.

6.3.1 Voice in the 3GPP Release 4 Network

The 3GPP Release 4 standards provide a means for operators to migrate the Release 1999 circuit-switched domain to an IP-based core network infrastructure, independently of the UTRAN and GPRS networks. The Release 4 standards include other changes and new features which do not directly affect the voice service, and hence are not described here.

An overview of the Release 4 network is shown in Fig 6.2.

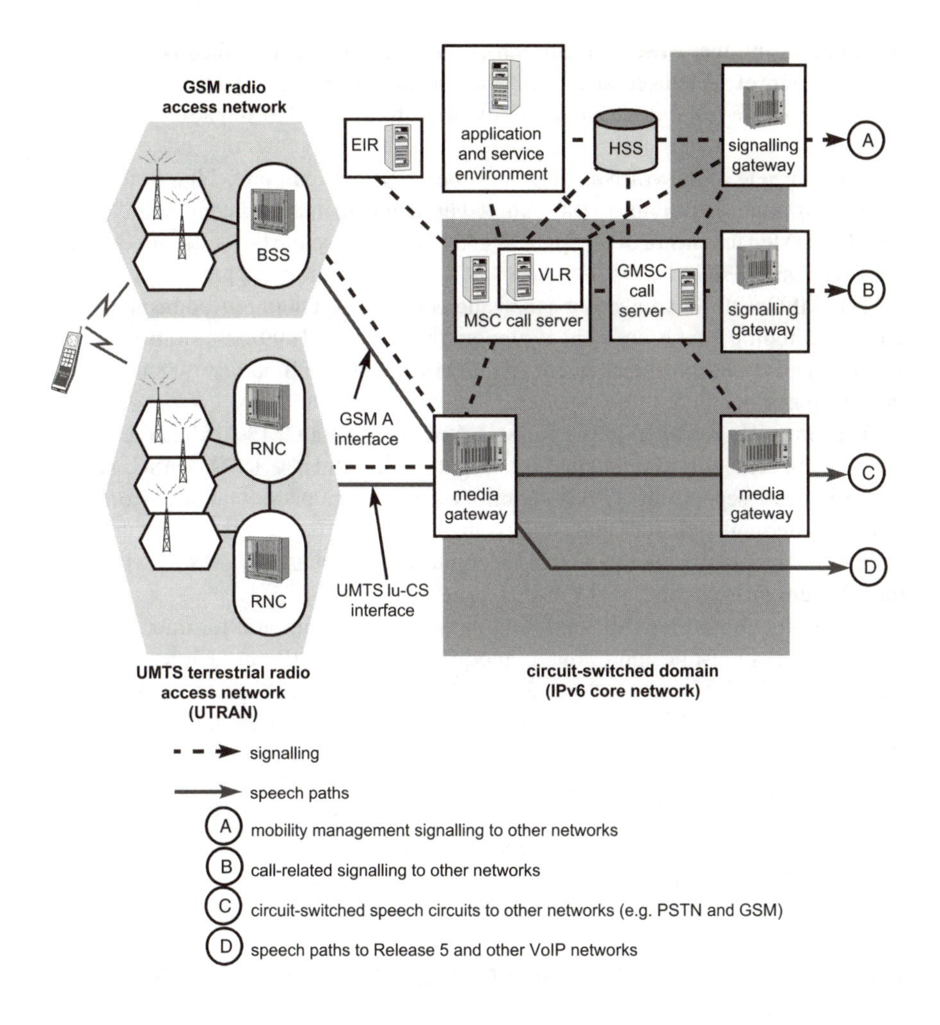

Fig 6.2 3GPP Release 4 network overview.

The Release 4 circuit-switched domain connects to the UTRAN via the Iu-CS interface, which supports the same speech transport and signalling protocols as in Release 1999. GSM radio networks can also be connected via the GSM A interface. However, rather than terminating on an MSC, the Iu-CS interface (and similarly the GSM A interface) terminates on a media gateway (MGW) in the circuit-switched domain.

The media gateway relays the signalling and voice circuits from the ATM transport used on the Iu-CS interface to an IP-based core network. The signalling is carried to the MSC call servers in an appropriate Internet protocol such as the transmission control protocol (TCP) (see RFC 761 [4]) or stream control transmission protocol (SCTP). Speech paths in the Iu-CS interface are converted into a stream of IP packets using the user datagram protocol (UDP), retaining the AMR framing (a fuller description of how speech is carried in IP packets is given in section 6.4.3).

The mapping of the voice streams and speech path circuit-switching in the media gateway is controlled by the MSC call server, using an appropriate protocol such as the H.248/Megaco protocol (see RFC 2885 [4] and Rosen [5]), jointly produced by the ITU-T and IETF.

The MSC call servers can be considered as MSCs without embedded voice circuit-switching capability. They support the Release 1999 call control, service features and mobility management of an MSC. MSC call servers are more fully described in Chapter 5.

The GMSC is replaced in the Release 4 network by a GMSC call server, which performs the call control and HLR interrogation of a Release 1999 GMSC. As with the MSC call server, the GMSC controls a media gateway which performs the circuit-switching function and relays the voice IP packets into the transport technology being used on the interconnect interface to other networks — which may be TDM circuit-switched, ATM AAL2, or IP.

The protocol used for call signalling in the core network and for interconnect to other networks must be appropriately modified to support IP-based voice paths. For example, the ITU-T bearer-independent call control (BICC) [6] protocol can be used instead of the ITU-T SS7 integrated services user part (ISUP). A signalling gateway (SGW) is used to relay and police the interconnect signalling protocol, mapping the signalling transport protocols used on each side of the gateway.

Using this design, the Release 4 networks are capable of supporting the Release 1999 voice service with minimal enhancement to the network and little, if any, impact on the end user.

6.4 End-to-End Voice and Multimedia on IP — 3GPP Release 5

The migration to a Release 4 voice network enables the use of the Internet protocol within the core network to provide the same voice services as in Release 1999. While this does not provide any additional service features to the end user, it could prove to be more cost effective than a Release 1999 network. One reason for this is that Release 4 is a step towards the 3GPP Release 5 IP multimedia services network.

The Release 5 standards specify a voice and multimedia services network called the Internet protocol multimedia (IM) subsystem, or IM subsystem for short. The IM subsystem relies on a managed GPRS and core IP network that is enabled to provide the quality of service needed for voice and multimedia services.

The main reasons for the introduction of the IM subsystem are to enable new services and to reduce cost. The IM architecture uses IP and the other protocols standardised by the Internet Engineering Task Force (IETF) as interfaces to component 'building blocks' of the Release 5 network.

These protocols provide a very adaptable suite of technologies for building packet-based networks and services, and the growth in the use of these protocols and associated networking equipment over the last decade has resulted in considerable cost reductions. However, while the IETF protocols can be adopted to provide many of the functions of the IM subsystem, each UMTS service has specific requirements that impact on the overall design of the network and the detailed information carried within the protocols. Therefore, to determine the IM network and protocol design, the services to be supported must be understood.

Examples of the services that will be supported in Release 5 by the IM subsystem are:

- voice telephony;
- real-time interactive games;
- videotelephony;
- instant messaging;
- emergency calls;
- multimedia conferencing.

These services tend to share a number of characteristics — they are generally a conversational session between two or more parties requiring some degree of real-time interactivity. The real-time aspects of the service can be described in terms of the quality of service of the transport (such as transmission delay or packet jitter) and of the session (or call) control, such as time to establish the session.

To meet the interactive needs of these services, the GPRS network provides quality-of-service levels — for example, by operating at low levels of network utilisation or by employing mechanisms such as diffserv (see RFCs 2474 and 2475 [4]). Additionally, IP version 6 (IPv6 — see RFC 2460 [4]) has been recommended as the transport protocol to be used for the IM subsystem, since this has a number of features that are beneficial to UMTS networks (such as a large address space, support for packet prioritisation, and easier manageability).

The IM subsystem has four important roles in meeting the requirements of services:

- it enables users and applications to control the sessions and calls between multiple parties, for example to establish, maintain, modify and terminate calls[1];

[1] Strictly speaking, sessions and calls are different (see RFC 2543 [4] for a definition of each). However, for the purposes of this chapter the term 'call' is used to refer to simple cases where calls and sessions can be considered the same, for example, in the case of a point-to-point voice telephony call.

- it controls and supports network resources (such as media gateways and GPRS gateway support nodes (GGSNs), multimedia resource functions (MRF), and the core IP network) to provide the functionality, security and quality required for the call;
- it provides for registration of users on the 'home' and 'roamed to' networks, so that users may access their services from any UMTS network;
- it generates call detail records (CDRs), for example containing information on time, duration, volume of data sent/received, and the call participants — the CDRs, together with records from the GPRS network on the data volumes transmitted and received are used for charging purposes.

An overview of the IM subsystem and its relationship with the GPRS packet-switched domain is shown in Fig 6.3. The purpose of each of the functional entities is more fully described in Chapter 5.

The IM subsystem architecture complements the voice over IP (VoIP) protocols and architectures developed by the IETF [4], ETSI TIPHON [7] and ITU-T Study Group 16 [3], although these were primarily developed for fixed IP networks (see also Swale [8]). Supporting VoIP in a mobile and wireless environment raises a number of additional requirements, which are being addressed by the 3GPP group. In particular, they include:

- the ability of the network to hand over a call (signalling and speech paths) from one radio base-station (node B) to another, without perceivable loss of speech quality, for example as a user moves between radio base-stations;
- the ability of the network to cope with the additional delays imposed on the speech path due to radio access and use of AMR coding, without perceivable loss of speech quality;
- the ability of the network to allow users to roam to another operator's network (a visited network), and still receive service;
- the ability of the user's home network to control the voice service of a roaming user on a visited network — the users's home network is the one to which the user is subscribed.

The first two points are addressed by the mechanisms used to transport IP packets carrying speech and IP packets carrying signalling over the GPRS network and IM subsystem core IP network. The subsequent points are addressed by the registration, discovery and call control procedures of the IM subsystem.

6.4.1 Overview of VoIP in 3GPP Release 5

In common with fixed network VoIP, digitised speech from each user is carried in IP packets between one user's terminal equipment and another by an IP network.

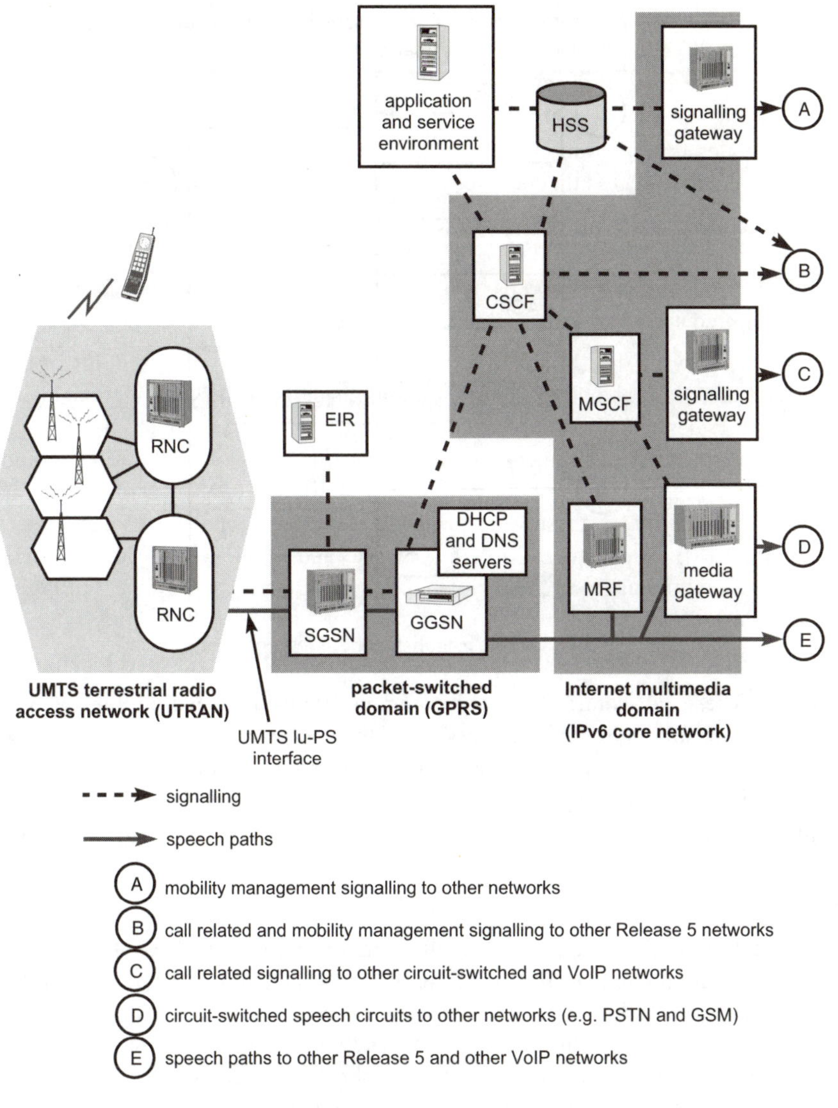

Fig 6.3 3GPP Release 5 network overview.

The path that these packets take through the network is referred to as the speech path. Unlike a circuit-switched environment, the packets may individually take different routes through the IP core network to a common exit point of the IP core network, rather than be forced along a specific circuit. However, in reality, it is likely that the packets will follow the same route through the network if the network is not congested. To establish a speech path, and synchronise the users and their equipment, call-control functionality is programmed into the user's equipment and

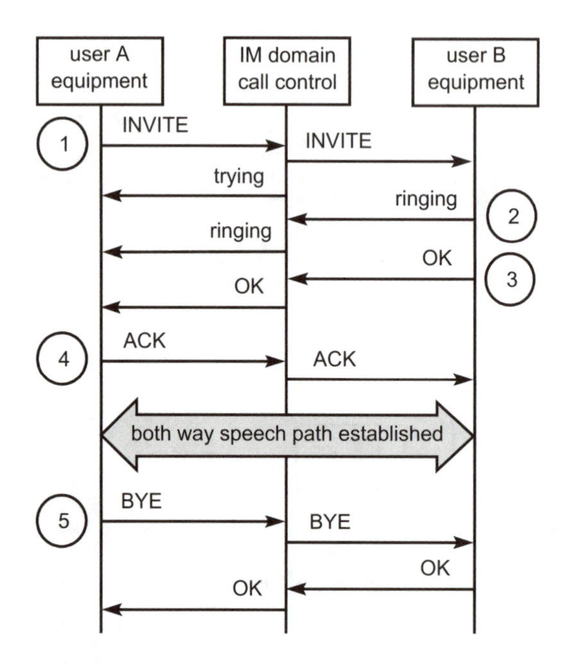

Fig 6.4 Simple establishment of a speech path.

network. These call-control functions communicate using signalling messages. For example, the call control enables passing of the end-point addresses for the speech paths on the user's equipment and the negotiation of the network and user equipment resources needed for the call, such as codecs and the quality of service required. Figure 6.4 shows a simple call establishment to create a VoIP speech path, which is described below.

- Invite (1)

  The calling user (A) initiates the call by inviting the called user (B) to the call. This invitation contains a name representing the called user (either similar to an Internet e-mail address or a telephone number), a description of the call (e.g. codec to be used) and the address of the endpoint of the speech path on user A's equipment (e.g. a telephone). A call control entity in the IM subsystem receives this invitation, and confirms back to user A's telephone that it is trying to contact user B's telephone. The call control entity then performs a database look-up to translate user B's name to an address to which it can route the invitation. On resolving the address, the IM subsystem call control routes the invitation on to user B's telephone.

- Alerting (2)

  On receiving the invitation, user B's telephone alerts user B of the incoming call, and informs user A via the IM subsystem that the called telephone is ringing.

- Answer (3)

  When user B answers, the telephone accepts the call by sending an OK back to user A's telephone via the IM subsystem. This message contains the address on user B's telephone on which the speech path should terminate, as well as the agreed call description.

- Acknowledge (4)

  User A's telephone acknowledges acceptance of the call, and the speech path is established — both telephones now know each other's address and are able to send speech packets to each other.

- Release (5)

  When the users have finished talking to each other, the call is cleared, for example by user A's telephone sending a BYE to user B's telephone via the IM subsystem. Both telephones then free up any resources allocated for the speech path, and user B's telephone confirms that the speech path has cleared by sending an OK back to user A's telephone via the IM subsystem.

6.4.2 Signalling

The signalling protocol for registration and call control in the IM subsystem is based on the session initiation protocol (SIP) (see RFC 2543 [4]). In simple terms, the control of the call relates to inviting and synchronising the various participants in the call. It also enables the participants to describe and share information about the characteristics of the terminating equipment and the speech path between the users. This information is known as the session description, and could include, for example, the coder used for the speech and the bandwidth needed for the speech paths. SIP essentially provides the invitation and synchronisation of the participants, and it uses the session description protocol (SDP) (see RFC 2327 [4]) to describe the session.

Both of these protocols are standardised by the IETF, simple to use and program, text-based, and can be readily adapted to support a wide range of multimedia applications.

In the SIP protocol, users are addressed by a SIP uniform resource locator (URL), which has the form user_name@network_domain_name, where the user_name and network_domain_name are textual names (similar to an Internet e-mail address). PSTN telephone numbers may be textualised so that they can conform to this format, and thus allow users to be addressed from the PSTN (and vice versa).

The SIP URLs provide a flexible means of addressing that is well integrated with other Internet technologies. For example SIP URLs can be easily included in Web pages as hyperlinks, that when activated initiate a SIP session to that user.

The functional entity in the IM subsystem that performs the call control is known as the call state control function (CSCF). These have been classified into different types [6], and provide the functions of a stateful SIP proxy server, as defined in RFC 2543. Correspondingly, the user's equipment provides the functions of a SIP user agent, as defined in RFC 2543. SIP messages are usually transported by the transmission control protocol (TCP) (see RFC 761 [4]) or user datagram protocol (UDP) (see RFC 768 [4]). However, SIP is transport independent and other protocols, such as the stream control transmission protocol (SCTP) which runs over UDP, can be used to provide a higher level of quality than TCP.

The SIP itself includes reliability mechanisms that can be used if running over an unreliable transport, but these can be omitted if a reliable protocol transport such as TCP or SCTP is used. These protocols are carried over IP over the GPRS network and IM subsystem IP core network (see Fig 6.5).

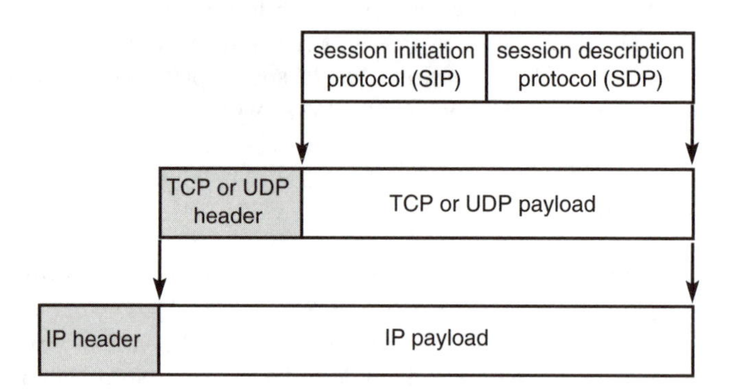

Fig 6.5 Transport of SIP and SDP in IP.

To send and receive SIP messages over the GPRS network, the user's equipment must establish a bi-directional packet data session with the IM subsystem for the signalling path. This is known as a packet data protocol (PDP) context activation and is a common GPRS procedure for establishing an IP data path between the user's terminal and the network. The signalling path is a separate PDP context to the speech path, and must be done before any SIP messages can be sent (e.g. for registration). The GPRS quality-of-service class for signalling is interactive, although the detailed parameters define the specific transport quality requirements (such as high priority, but lower sensitivity to delay and jitter). The establishment of the PDP context for signalling assigns an IP address to the mobile terminal and allocates bandwidth and the required quality of service over the UTRAN and GPRS network for the signalling. The assigned IP address, together with the SIP port number is used to address the SIP client in the mobile terminal. This IP address can also be used subsequently as the IP address for the speech paths.

When the mobile terminal is to be switched off or roams to another network, the PDP context for the signalling is deactivated.

One of the benefits of using GPRS to carry the signalling path is that the GPRS controls the handover of the signalling path as a user moves between the radio cells. It is therefore not essential that the IM subsystem be aware of the geographical location of a user. However, some voice services may require location information (for example, to restrict the user to certain cellular areas or for emergency services). In these cases, the IM subsystem will need to obtain location information from the GPRS network or home subscriber server (HSS).

It should be recognised that SIP only supports the call control procedures for the establishment of the speech path. The allocation of the actual bandwidth and quality of service needed to provide the IP transport for the speech packets over the UTRAN and GPRS network is requested by the user's equipment as additional PDP contexts using the GPRS protocols. Similarly, the control of quality-of-service mechanisms in the core IP network is independent of the call-control procedures, and instead relies on other solutions (such as diffserv).

Within the IM subsystem, additional protocols are used between network elements in order to provide the full voice service. These include a mobility management protocol between the CSCFs and the HSS (this could be MAP [2] or LDAP (see RFC 2251 [4])), and media gateway control protocol, such as the H.248/Megaco protocol (see RFC 2885 [4]), jointly produced by the ITU-T and IETF.

6.4.3 Transport of Speech Packets

So that speech may be sent and received in IP packets, the user's actual speech is sampled by the user's equipment and coded for transmission (e.g. using the AMR coder). Once a certain number of samples have been taken, usually between 10 ms and 40 ms, the coded samples are packetised and sent to the network. The time taken to packetise the speech samples adds considerable delay to the speech path and can necessitate echo cancellation devices within the terminal equipment or network. However, it is inefficient to simply send smaller packets of speech samples, since this increases the bandwidth needed and requires the IP routers to route more speech packets. In addition to the end-to-end transmission delay, additional delay due to packet jitter is encountered at the termination of the speech path where the packets have to be buffered so that the digitised speech can be synchronised before it is played out.

The speech packets are transported between users' equipment in UDP/IP packets by the GPRS network and IM subsystem core IP network. A framing protocol is required for the speech samples, e.g. to synchronise samples and control the sampling rate. The IETF protocols for framing voice and multimedia are the real-time transport protocol (RTP) (see RFC 1889 [4]) and RTP control protocol (RTCP), which are carried in UDP/IP packets. This is shown in Fig 6.6.

Currently, as RTP and RTCP do not support rate-controlled codecs such as AMR, another framing protocol such as the Iu user plane protocol (IuUP) [2], that is

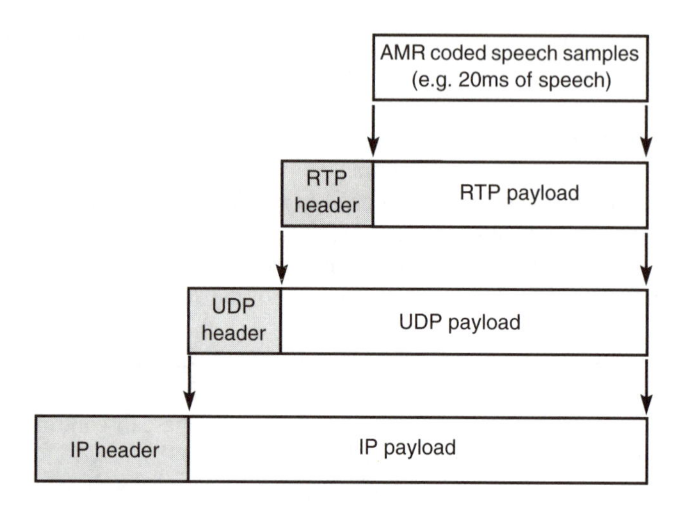

Fig 6.6 Transport of speech in IP.

used to frame the speech on the Iu interface in Release 1999, could be used instead of RTP and RTCP. However, for the purposes of this chapter, RTP and RTCP are used to illustrate the framing and transport of speech and multimedia in the IM subsystem.

For the user's equipment to be able to send and receive speech packets to and from the IM subsystem, it must activate a bi-directional PDP context between itself and the IM subsystem. This allocates bandwidth and the required quality of service over the UTRAN and GPRS network for the transport of speech packets. The entry point to the IM subsystem will contain firewalls for security and prevention of denial of service attacks, and these may also be controlled dynamically by the call control, on a call-by-call basis, to prevent speech packets being sent or received before the call is established. Deactivating the appropriate GPRS PDP context disconnects the speech path between the user's equipment and the IM subsystem.

To provide a higher quality of service than 'best-effort', the GPRS network specifies a conversational class of service that prioritises speech packets for low delay and low jitter. Similarly, mechanisms such as diffserv or the resource reservation protocol (RSVP) (see RFC 2205 [4]) may be used within the IM subsystem core IP network to provide a high-quality service.

The AMR coder is the default codec that all Release 5 terminals must support, although other codecs may be supported. As transcoder-free operation is supported by the call control signalling, AMR coding of the speech can be used end-to-end between the users' items of equipment, without the need to transcode to another standard. However, with AMR coding of speech at 12.2 kbit/s^{-1}, a 20 ms sample of speech results in an RTP speech payload that is roughly half the size of the combined IP, UDP and RTP packet headers. This makes for a very inefficient use of bandwidth, especially in the costly radio access. Increasing the sample size reduces the problem, but increases the end-to-end delay of the speech packets, as well as increasing the likelihood of packet loss on the radio interface.

One solution to this is to perform header compression between the user's equipment and the UTRAN, where the bandwidth is most expensive. This could theoretically bring the overhead of the IP, UDP and RTP headers down to less than 10% (not including the overhead of the lower layer GPRS and UTRAN protocols). Another possible solution is to transport the speech from the user's equipment through the UTRAN using AAL2, and to packetise the speech into IP payloads at the node B or RNC. However, both of these solutions require that speech be transported differently to other real-time and best-effort services that can be sent in uncompressed packets all the way to the user's equipment. For example, media such as video have transport-quality requirements that are similar to voice, but the higher bandwidth nature means that the payload-to-header ratio is much greater, and hence less wasteful of bandwidth.

As with the signalling path, a benefit of using GPRS to carry the speech paths is that the GPRS controls the handover of the speech paths as a user moves between the radio cells. However, this does require that the GPRS handover procedures be enhanced to ensure that the quality of service required for voice is met throughout the handover.

6.4.4 Roaming, Registration and Discovery

One of the main benefits of current GSM networks is the ability for the user to make and receive calls while travelling abroad. To provide such a benefit, the user must have the capability to be able to connect to a network that is controlled by an operator other than that to which they are subscribed. This benefit is also an essential feature of the Release 5 standards, although additional procedures are required to provide a roaming capability for the IM subsystem as the user's voice service is controlled directly by the user's home network. This is the main difference when roaming in an IM subsystem compared to the earlier UMTS releases and GSM, where the visited network always controls a roaming user's voice service. The IM subsystem allows some services to be provided by the visited network to roaming users, but these are not user-specific. For clarity, visited network control is not described further in this chapter.

Chapter 5 shows how the call is controlled by an entity known as a serving call state control function (S-CSCF). The S-CSCF is located in the user's home network, and hence the control procedures are referred to as home network control.

In order that the user can make and receive calls, the user equipment (UE) has to be registered with an S-CSCF. The registration procedure happens immediately after the user's equipment is switched on. Once registered, users can make and receive IM subsystem calls until they deregister. Before the registration procedure can take place, the user equipment has to connect to the network and discover an entry point into the IM subsystem. This entry point is the proxy call state control function (P-CSCF), and it provides a simple, generic call control function as well as potentially providing a firewall capability to ensure security of the IM subsystem.

The P-CSCF always resides in the network to which the UE is connected, and therefore the procedure for discovery of the P-CSCF is the same, irrespective of whether the user is roaming or not.

The procedure for discovery relies on GPRS signalling with the use of the IETF dynamic host configuration protocol (DHCP) (see RFC 2131 [4]) and domain name system (DNS) (see RFC 1035 [4]) protocols. The idea of the procedure is for any UE to be able to attach to a GPRS network, and be provided with an IP address of the P-CSCF. All SIP-based signalling from the UE then goes via the P-CSCF which is responsible for routing the messages on to the S-CSCF. Figure 6.7 shows the sequence of events in the discovery procedure.

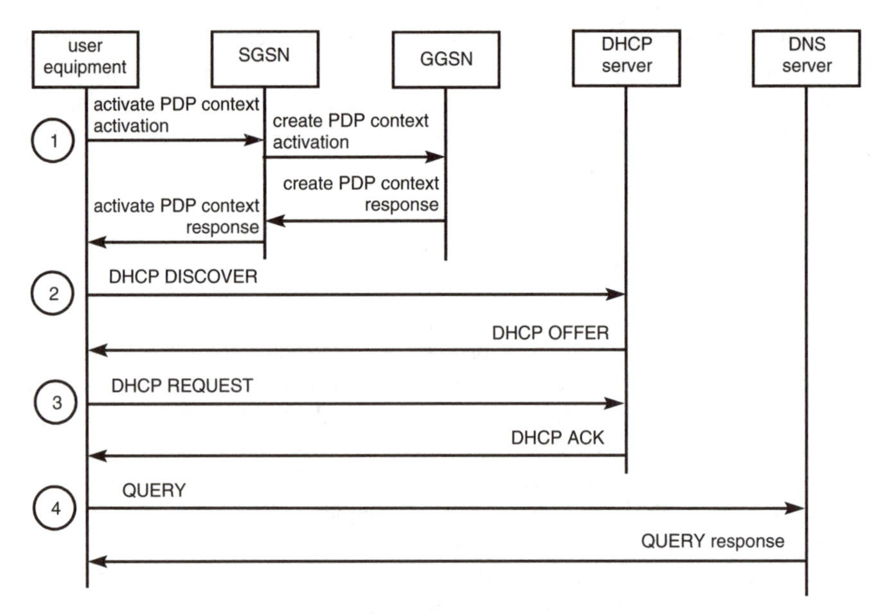

Fig 6.7 Discovery message sequence.

The sequence of events that make up the discovery procedure is described below.

- PDP context activation (1)

  The UE activates a PDP context to the GPRS network, which will be used for the discovery procedures, and later for the IM subsystem registration and call control procedures using SIP. To achieve this, the UE sends an activate PDP context activation request to the SGSN. Upon receipt of the request, the SGSN sends a create PDP context activation request to the GGSN. If the GGSN is able to establish a PDP context (e.g. after checking that the UE has the necessary permission), it creates a PDP context response to the SGSN, which in turn replies to the UE with an activate PDP context response. This is a standard GPRS procedure, although the details, such as the GPRS address point name used and the nature of the PDP address returned, may be specific to the discovery procedure.

- DHCP discovery (2)

  The UE broadcasts a DHCP DISCOVER message to the network. Upon receiving this message the DHCP server can respond with a DCHP OFFER message or it may not respond at all. If the DHCP server decides to respond it broadcasts the DHCP OFFER message with a specified available IP address. It should be noted that at this stage there is no agreement of an assignment between the DHCP server and the UE. The UE may receive more than one DHCP OFFER response (if more than one DHCP server responds) and therefore will have to choose one.

- DHCP request (3)

  Using the IP address received within the DHCP OFFER response, the UE broadcasts a DHCP REQUEST message containing the chosen IP address to the server(s). Each server checks the returned IP address. If it does not match, the server considers it as an implicit decline. However, the selected DHCP server sends a DHC PACK to the UE.

- DNS query (4)

  The UE sends a DNS QUERY to the DNS server for resolution of the predefined name for P-CSCFs to an IP address. The DNS server replies to the UE with a QUERY response containing the IP address of an appropriate P-CSCF.

  On disconnection of the UE, such as just before the device is turned off, the IP address can be released back to the DHCP server and the signalling PDP context can be deactivated. Now that the UE has knowledge of the proxy CSCF address, the registration procedure can take place in order that an S-CSCF can be selected. Figure 6.8 shows the functional entities involved in registration for home network control, and Fig 6.9 shows the message sequence required. If a user is connected to the home network rather than a visited network, the visited I-CSCF is not required and the P-CSCF will be in the home network.

After the UE has obtained a signalling path through the GPRS network, it can perform the IM registration. Signalling based on SIP is used to perform the registration between the UE and the CSCFs. The protocol between the CSCFs and the HSS is as yet undefined, but is represented in this chapter by information flows prefixed by the letters Cx (since this is the Cx reference point in the architecture). IM subsystem registration requires the following steps.

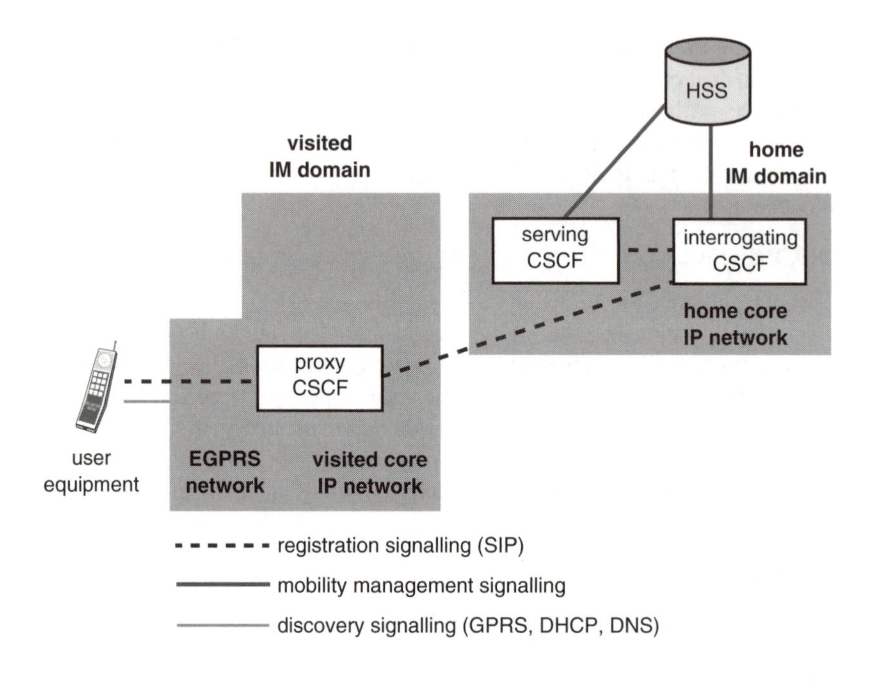

Fig 6.8 Functional entities for registration.

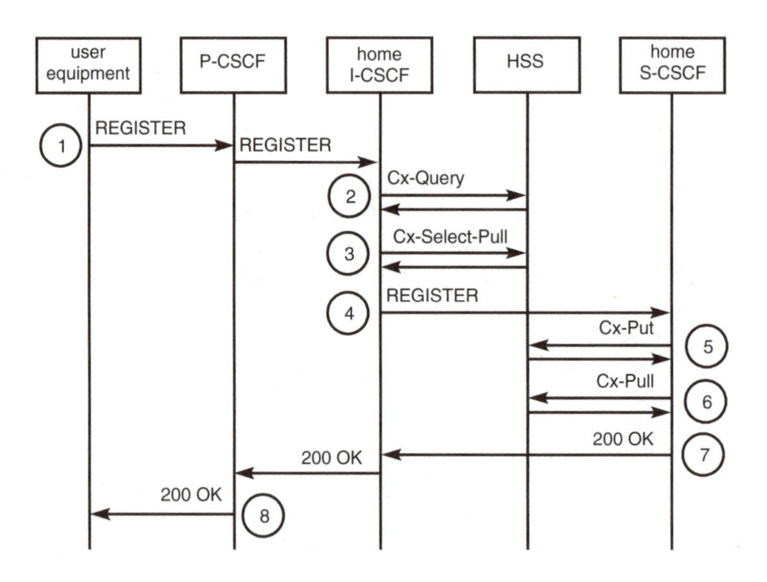

Fig 6.9 Message sequences for registration.

- Register (1)

  The UE sends a REGISTER message to the P-CSCF. This message contains the subscriber identity and the domain name of the home network. Upon receipt of the REGISTER, the P-CSCF examines the home domain name to determine the entry point to the home network. This entry point is an interrogating CSCF (I-CSCF), which provides policing of the SIP interface to other networks and interrogation of the home subscriber server. The P-CSCF forwards the REGISTER message on to the I-CSCF in the home network, adding the name of the P-CSCF, a visited network contact point name, and the visited network capabilities. A name-address resolution mechanism is utilised in order to determine the address of the home network from the home domain name.

- HSS Registration (2)

  When the I-CSCF receives the REGISTER message, it queries the HSS by sending a Cx-Query containing the parameters of the REGISTER message. The HSS checks whether the user is already registered, and carries out any necessary authentication. The HSS then issues a response to the I-CSCF indicating whether the registration and authentication were successful.

- S-CSCF capability identification (3)

  The I-CSCF then sends a Cx-Select-Pull to the HSS to request the information related to the S-CSCF capabilities required by the user. The HSS responds with the necessary information on the required S-CSCF capabilities to the I-CSCF.

- S-CSCF selection (4)

  The I-CSCF then selects an appropriate S-CSCF based on the capabilities required, and forwards the REGISTER message on to the S-CSCF.

- S-CSCF/user association (5)

  On receiving the REGISTER, the S-CSCF associates the subscriber and the S-CSCF name in the HSS using the Cx-Put, which is acknowledged by the HSS.

- User profile retrieval (6)

  The S-CSCF then uses the Cx-Pull request/response to retrieve the subscriber's profile for the user from the HSS, which it then stores locally. The S-CSCF also stores the name of the P-CSCF.

- Serving contact name determination (7)

  The S-CSCF determines whether the entry point into the home network from the P-CSCF should be the S-CSCF directly, or an I-CSCF for network hiding purposes. The S-CSCF then returns a 200 OK message with this information to the home I-CSCF. The home I-CSCF forwards the 200 OK to the P-CSCF, and then releases all knowledge of the registration information for that user.

- Registration completion (8)

  On receiving the 200 OK message, the P-CSCF stores the serving network contact name, before sending the 200 OK to the UE and completing the registration procedure.

The user is now registered with an S-CSCF in the home network and is able to make and receive voice and multimedia calls with the IM subsystem, irrespective of the location.

Deregistration and re-registration procedures are also defined, for example, to remove the registration of the user on an S-CSCF.

6.4.5 Control of Voice (and Multimedia) Calls

Once the user is registered with an S-CSCF, voice and multimedia calls may be made to other users.

The S-CSCF provides the main point of control of the call and any supplementary or advanced service features for that user. SIP signalling between the user equipment and the S-CSCF is routed via a P-CSCF, which provides a (secure) entry point to the IM subsystem and a point of flexibility for routing SIP messages to S-CSCFs.

Each user will be registered with an S-CSCF, so that a simple voice call between two users will usually require two S-CSCFs to communicate (i.e. one for each user). Additionally, an I-CSCF is required in order to interrogate the HSS to find the S-CSCF on which the called user is registered. Figure 6.10 shows the main functional entities involved in the control of voice calls between two mobile users on a Release 5 network.

For simplicity, this scenario assumes that both users are connected to, and registered on, their home network (i.e. they are not roaming). However, the sequence of events is similar for roaming users.

An IM subsystem call comprises the following five distinct phases.

- Call invitation

  The calling user invites the called user to participate in a call. This is supported by the SIP INVITE method and the 100 trying provisional response.

- Resource reservation

  The resources (such as the GPRS and UTRAN bandwidth and codecs to be used) are negotiated and reserved, for example, so that early tones and announcements can be played, and that transport for the speech path is available when the called user answers. The requirements for this phase are not yet fully supported by SIP.

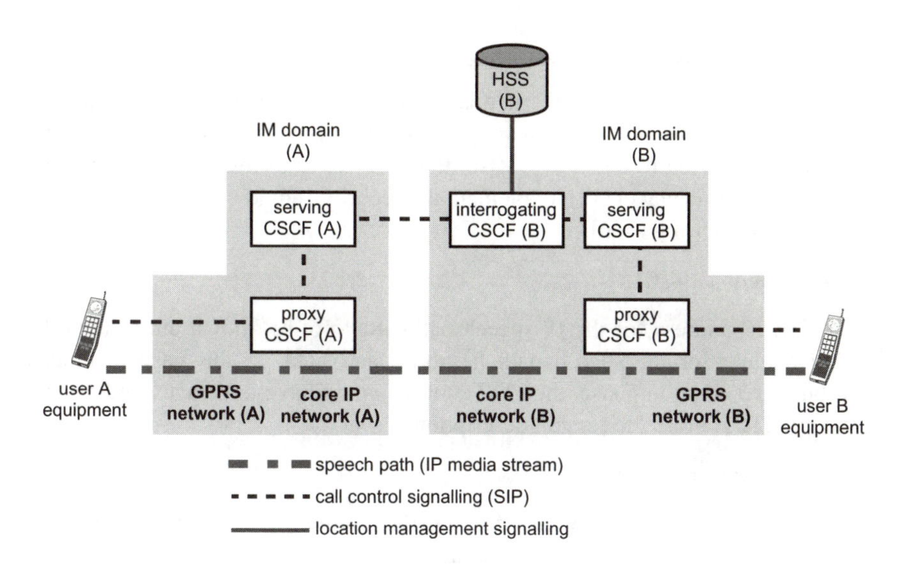

Fig 6.10 Functional entities in a Release 5 mobile-to-mobile call.

- Call alerting

  The called user is alerted to the incoming call. Support for informing the calling user of this event is provided by the SIP 180 ringing provisional response.

- Call connection

  The called user answers, the speech path is connected and charging begins. This is supported in SIP by the 200 OK final response and the ACK method.

- Call termination

  The call and speech path is cleared by one of the users. This is supported in SIP by the BYE method and the 200 OK final response.

The resource reservation phase of the call is necessary in the mobile environment for a number of reasons, which include:

- Path establishment prior to ringing

  The establishment of the PDP contexts for transport of the speech path should occur prior to the called user's telephone ringing. While this may not need to be the case for all multimedia services, there is a user expectation that when a ringing telephone is answered, a speech path will be in place. Given the scarcity of bandwidth on the radio interface, if reservation is not performed early, then in some cases a ringing telephone could be answered only for the users to find that there is no speech path available.

- Quality of service

  The service may require that the quality of service of the speech channels be established end-to-end, using a protocol such as the IETF RSVP, or that the speech paths need to be secure. If so, the procedures to reserve the appropriate quality-of-service level or to implement the security should occur prior to the call ringing and alerting user B.

- PDP content activation

  The UE IP address for the IP speech paths may not be known until the GPRS PDP context for the speech path has been activated. In this case, without a resource reservation phase, the PDP context would have to be activated prior to the INVITE, before the session description is sent.

- Tone/announcement provision

  Without early reservation of the speech path, it is not possible for the network or end equipment to provide tones or announcements in the speech path back to the calling user prior to the call being answered (such as ringing tone or busy tone).

Figure 6.11 illustrates the SIP signalling flows for a simple mobile-to-mobile call with a resource reservation phase, based on the scenario in Fig 6.10. It assumes that the underlying GPRS and IP core network provide the necessary quality of service for the speech paths.

The message sequences in Fig 6.11 are described below.

- Invite — originating network (1)

  User A initiates the call by sending an INVITE message to the P-CSCF, which contains the names of the calling and called users as SIP URLs. The session description part of the message includes the IP address of user A's UE (if known) and an initial description of the session (e.g. AMR coded speech using RTP, with the UDP port number). This description may include options, such as a range of codecs that could be used. Additionally, the session description indicates that the reservation of the speech path IP transport and quality of service is a mandatory pre-condition to ringing.

  The P-CSCF confirms receipt of the INVITE by replying with a 100 trying message, and forwards the INVITE on to the S-CSCF, adding the name of user A's P-CSCF to the message. This allows tracing of the signalling route back through the network.

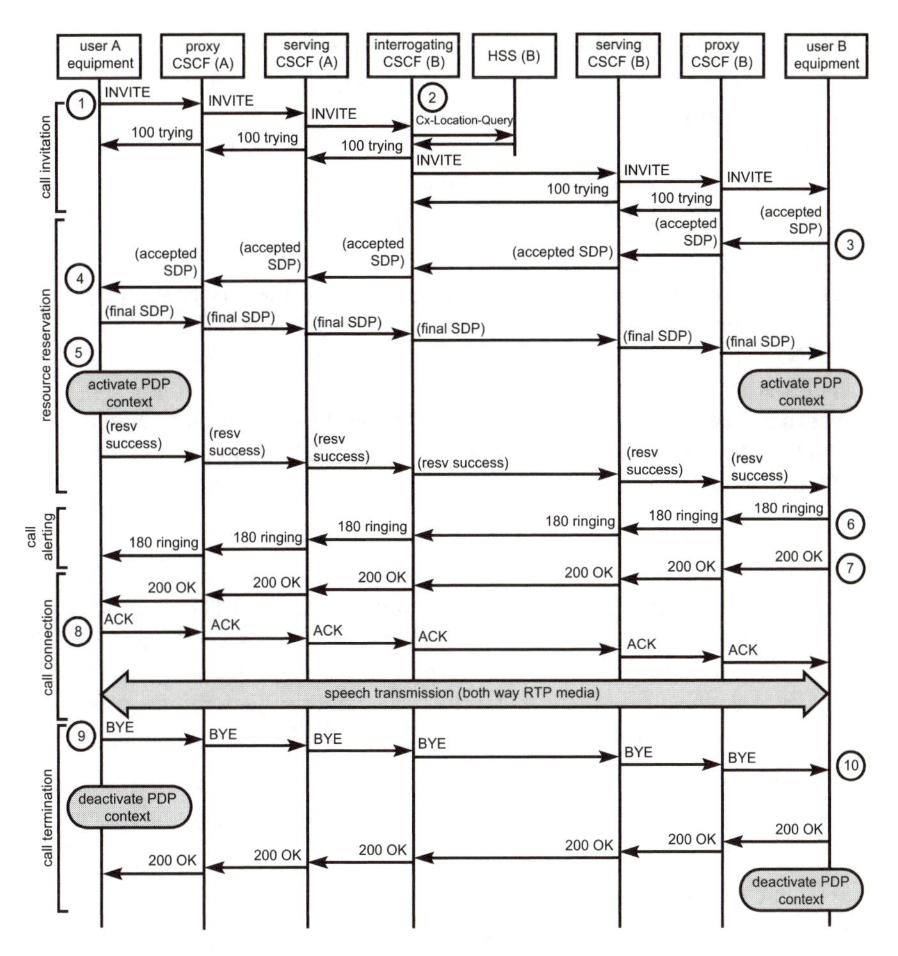

Fig 6.11 Signalling message sequences for a simple mobile-to-mobile call.

The S-CSCF confirms receipt of the INVITE by replying with a 100 trying message, then validates the service profile for user A and invokes any necessary service features (for example, outgoing call-barring). The S-CSCF then determines a SIP entry point for user B's home network from the SIP URL for user B, for example by performing a DNS query. The SIP entry point to user B's home network will usually be an I-CSCF. User A's SCSF then sends the INVITE on to user B's I-CSCF, adding the name of user A's S-CSCF to the message.

- Invite — terminating network (2)

  On receiving the INVITE, the I-CSCF interrogates the HSS with a Cx-Location-Query to determine the address of user B's S-CSCF. It also confirms receipt of the INVITE from user A's S-CSCF by replying with a 100 trying message. Once the HSS responds with the address of the S-CSCF on which user B is registered, the I-CSCF forwards the INVITE on to that S-CSCF, adding its name to the message.

  User B's S-CSCF receives the INVITE and validates the service profile for user B. It then invokes any necessary service features for user B, before forwarding the INVITE on to user B's P-CSCF adding its name to the message. The S-CSCF confirms receipt of the INVITE by replying to the I-CSCF with a 100 trying message. The P-CSCF receives the INVITE and forwards it on to user B's UE. The P-CSCF confirms receipt of the INVITE by replying to the S-CSCF with a 100 trying message.

- Session description negotiation (3)

  User B's equipment accepts the call invitation, but does not alert user B at this stage. Instead, it checks the session description in the INVITE and determines whether it is acceptable. If so, it responds with a message[2] to the P-CSCF, indicating the acceptable parts of the session description (for example, reducing the list of optional codecs). User B's P-CSCF authorises any resources required, and then forwards the message to user A's P-CSCF along the signalling path via the CSCFs.

  User A's P-CSCF also authorises any resources required before forwarding the message to user A's UE.

- Session description agreement (4)

  User A's UE decides the final session description, for example deciding which codec to use from the acceptable options. This final session description is then sent in a message[2] that traverses the signalling path to user B's UE.

- Reservation of speech path resources (5)

  Once the final session description is agreed, both UEs can each activate a GPRS PDP context for the speech path. This stage could also include the reservation of the necessary quality of service for the speech paths (such as bandwidth and maximum delay) on an end-to-end basis.

  When user A's UE has completed the PDP context activation and any other resource reservation, it sends a message[2] indicating that resource reservation was successful to user B's UE, via the signalling path.

---

[2] It should be noted that the SIP message to convey this information has not yet been agreed as part of the SIP protocol specification.

- Alerting (6)

  Once user B's UE has activated its PDP context and reserved any other resources, it waits for the resource reservation success message from user A's UE. When this is received, it alerts user B, for example by ringing. It indicates this back to the P-CSCF using a 180 ringing message, which is sent back via the signalling path to user A's UE. User A's UE will then provide an indication of this back to user A, such as a locally generated ringing tone.

- Answer (7)

  User B answers the call. User B's UE sends a 200 OK message via the signalling path to user A's UE. On receiving the 200 OK, both P-CSCFs commit the resources that have been reserved. It is likely that the P-CSCFs will have some control over the IM sub-system speech-path entry points (firewalls), and not permit the speech packets through until this stage, or on receipt of the subsequent ACK (the choice may be service dependent). This control could also be the point at which the call charging commences.

- Acknowledgement (8)

  User A's UE acknowledges the establishment of the call by sending an ACK, which traverses the signalling path back to user B's UE. The UEs are now able to send IP speech packets to each other.

- Release — originating network (9)

  To release the call, user A's UE sends a BYE message to user B's UE via the signalling path, and deactivates its PDP context for the speech path. At this point, the P-CSCF may close the IM subsystem speech-path entry point to further traffic and cease charging.

- Release — terminating network (10)

  User B's UE responds by deactivating its PDP context for the speech path and acknowledging the BYE with a 200 OK. This traverses the signalling path back to user A's UE, releasing each of the CSCFs from the call.

6.5 Interworking 3GPP Release 5 with Other Networks

Connection to circuit-switched networks, such as the PSTN, GSM and 3GPP Release 1999 networks, requires interworking at both the speech path level and the signalling level. Media gateways (MGWs) are included in the IM subsystem to terminate the IP transported speech paths and convert them to circuit-switched TDM, transcoding if necessary. The circuit-switched networks generally use the ITU-T SS7 integrated services user part (ISUP) to control calls. Signalling gateways

(SGW) in the IM subsystem map between the message transfer part levels of SS7 and the SIP transport protocol (e.g. TCP/IP) used in the IM subsystem.

It is not possible to simply map ISUP signalling messages into SIP messages, since the service context of the messages must be known. A media gateway control function (MGCF) is used to perform the mapping of the IM subsystem voice service (and SIP signalling) to the voice service of the other network (e.g. PSTN voice service and ISUP signalling). The MGCF communicates with the S-CSCF or I-CSCF using SIP. The MGCF also controls the MGW, for example using the H.248/ Megaco protocol [5].

Interworking with other VoIP networks that are not compatible with 3GPP Release 5, such as those based on ITU-T Recommendation H.323, also requires signalling and media gateways in order to map any differences in lower layer protocols and police the IM subsystem. An MGCF is also needed to ensure appropriate mapping of the voice service between the networks. Figure 6.12 shows the functional entities involved in interworking 3GPP Release 5 voice calls with PSTN or GSM networks.

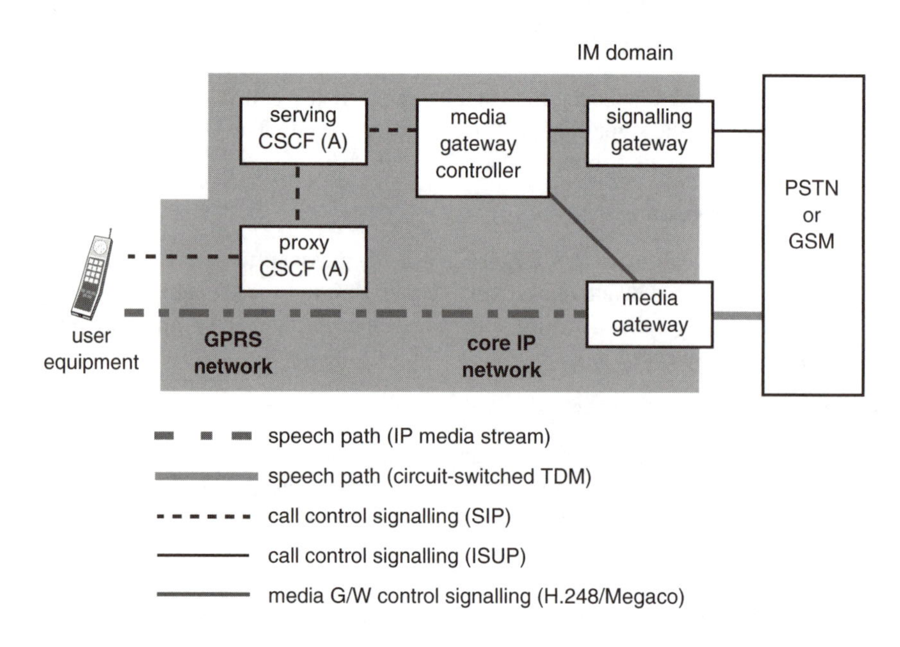

Fig 6.12 Functional entities in a Release 5 mobile-to-PSTN call.

Figure 6.13 illustrates the message sequences for a simple mobile originated voice call that terminates in the PSTN network. The message sequences are described below.

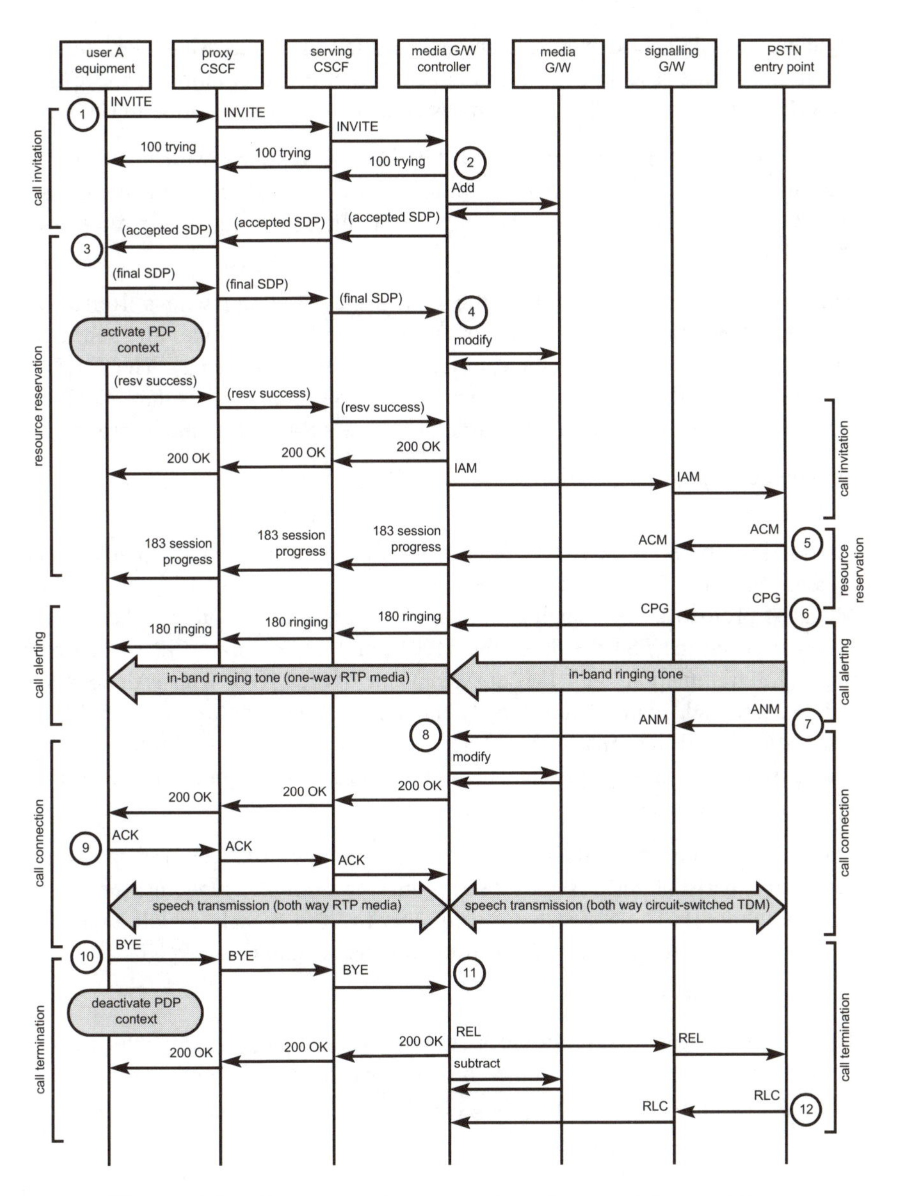

Fig 6.13 Signalling message sequences for a simple mobile-to-PSTN call.

- Invite (1)

  The UE of the mobile calling party (user A) initiates the call to the PSTN user by sending an INVITE to their P-CSCF, which contains an appropriate session description. The PSTN user (user B) is identified as such by the SIP URL, which contains the PSTN telephone number encoded into the SIP URL format. Additionally, the session description indicates that the reservation of the speech path IP transport and quality of service is a mandatory pre-condition to ringing.

The P-CSCF confirms receipt of the INVITE by replying with a 100 trying message, and forwards the INVITE on to the S-CSCF, adding the name of user A's P-CSCF to the message. This allows tracing of the signalling route back through the network.

The S-CSCF confirms receipt of the INVITE by replying with a 100 trying message, then validates the service profile for user A and invokes any necessary service features for user A (for example, outgoing call-barring). The S-CSCF then determines that the call is destined for the PSTN, and routes the INVITE to an appropriate MGCF, adding the name of user A's S-CSCF to the message.

- Session description negotiation (2)

  The MGCF initially responds to the S-CSCF with a 100 trying. It then checks the session description in the INVITE and determines whether it is acceptable. If so, it configures the MGW for the speech path[3] (for example using the H.248/ Megaco protocol), by seizing an already created circuit-switched trunk termination on the PSTN side of the MGW, and adding a new IP speech-path (e.g. RTP) termination to the IM subsystem side of the MGW. This is done by the 'add' command, which additionally creates a new context in the MGW associating the IP termination and PSTN termination. The PSTN termination is configured for both-way speech. The MGW returns a description of the ports to the MGCF in response.

  The MGCF then responds with a message[2] to the S-CSCF, indicating the acceptable parts of the session description (for example, reducing the list of optional codecs). The S-CSCF forwards this to the P-CSCF, which authorises any resources required, before forwarding the message to user A's UE.

- Session description agreement (3)

  The UE decides the final session description, for example, deciding which codec to use from the acceptable options. This final session description is then sent in a message[2] that traverses the signalling path to the MGCF.

- Reservation of speech-path resources (4)

  On receiving the finally agreed session description, the MGCF modifies the IP speech-path termination on the MGW to select the chosen codec and switch through the backward speech path from the PSTN so that the mobile user can hear tones and announcements from the PSTN.

---

[2] (see page 90)

[3] It is assumed that the MGW has already established a control relationship with the MGCF and the terminations on the TDM circuit-switched side have already been provisioned and configured.

Simultaneously, the UE activates a GPRS PDP context for the speech path. This stage could also include the reservation of the necessary quality of service for the speech path (such as bandwidth and maximum delay) through to the MGW. When it has completed this, the UE sends a message[2] indicating that resource reservation was successful to the MGCF, via the signalling path.

The MGCF now initiates the call establishment to the PSTN by sending an initial address message (IAM) to the signalling gateway (SGW). The SGW relays the IAM from the IP-based transport protocol (for example SCTP/UDP/IP) to the SS7 message transfer part, and on to the PSTN entry point (for example a PSTN gateway trunk exchange).

- PSTN acceptance (5)

  The PSTN accepts the call with an address complete message (ACM), which is sent back to the MGCF via the SGW. The MGCF may send a 183 session progress message back to the UE so that the mobile user may now hear any in-band tones and announcements from the PSTN. This indicates that one-way IP speech packets may be received and the address of the RTP termination on the MGW.

  The message follows the signalling path, and causes the P-CSCF to commit any necessary resources and control the IM subsystem IP speech-path entry point (e.g. firewall) from the GPRS network to allow the media to be played to the UE.

- PSTN alerting (6)

  The PSTN sends a call progress message (CPG) to the SGW, indicating that the called user's telephone is ringing. This may be accompanied by in-band ringing tone in the speech path back to user A. The SGW relays this message back to the MGCF. The MGCF sends a 180 ringing message to the S-CSCF, which is forwarded via the signalling path to the UE.

- PSTN answer (7)

  When the call is answered, the PSTN sends an answer message (ANM) to the SGW, which relays it back to the MGCF.

- MGW 2-way connection (8)

  At this point, the MGCF issues another modify command to change the IP speech-path termination in the MGW to allow both-way speech paths to be switched through. The MGCF then sends a 200 OK message back to user A's UE, with an indication that two-way media may be sent and received. This message follows the signalling path, and causes the P-CSCF to commit any necessary resources and control the IM subsystem IP speech-path entry point to allow both-way media. This is the point at which call charging commences.

- Acknowledgement (9)

  User A's UE receives the 200 OK. The call is now established and two-way speech can take place between the mobile and PSTN user. The UE acknowledges this by sending an ACK message back to the MGCF via the signalling path.

- Release — originating network (10)

  When the mobile user clears the call, the UE sends a BYE to the P-CSCF and deactivates the PDP context for the speech path. The P-CSCF forwards the BYE on to the MGCF via the S-CSCF.

- Release — terminating network (11)

  The MGCF releases the call into the PSTN with a release message (REL) and confirms the BYE by sending a 200 OK message back to the UE via the CSCFs on the signalling path, which each release the call in turn.

  The MGCF then clears the speech path in the MGW by issuing a subtract command to delete the terminations from the call context, and the call context itself. The MGW optionally responds by sending an audit report for the call to the MGCF that contains information such as the number of packets sent/received and the packet loss.

- Release confirmation (12)

  The release message sent to the PSTN is confirmed back to the MGCF (via the SGW) by a release complete message (RLC), which completes the release procedure.

6.6 Conclusions

Voice telephony is an essential service for many mobile network users, and one that must be supported by 3rd generation networks. This chapter has shown how the initial 3GPP UMTS standards have taken an evolutionary approach to providing a voice service compatible with GSM to maximise the benefits of the new radio access technologies. It has then described how a more innovative approach to providing voice and multimedia integration with the Internet protocols is being developed for the Release 4 and Release 5 standards.

References

1 Mehrotra, A.: '*GSM System Engineering*', Artech House (1997).

2 3GPP — http://www.3gpp.org

3 International Telecommunication Union — Telecommunication Standardization Sector — http://www.itu.int

4 Internet Engineering Task Force — http://www.ietf.org

5 Rosen, B.: '*VoIP gateways and the Megaco architecture*', BT Technol J, **19**(2), pp 66-76 (April 2001).

6 Knight, R. R., Norreys, S. E. and Harrison, J. R.: '*Bearer-independent call control*', BT Technol J, **19**(2), pp 77-88 (April 2001).

7 European Telecommunications Standards Institute project — TIPHON — http://www.etsi.org/tiphon

8 Swale, R. P. (Ed): '*Voice over IP*', BT Technol J, **19**(2) (April 2001).

7

3G SERVICE CONTROL

M D Cookson and D G Smith

7.1 Introduction

Service control is about controlling network resources to provide services in real time. This chapter concentrates on the opportunities for service control in future mobile, third generation (3G) networks. In particular, the focus is on the all-IP network, defined by the 3G Partnership Project (3GPP) [1] as the Release 4/5 network.

The subject is covered in three parts. Firstly, the business approach to services and the network architecture for the Release 4/5 network are discussed. Secondly, the intelligence strategy being used to direct the service control work is described. The last section brings together the preceding parts and describes how they are being pursued for service control in Release 4/5.

Service control encompasses the evolution of today's intelligent network, but also embraces the many new ways of providing real-time services. This includes terminal capabilities and IP technologies. It is used to refer to the software and signalling messages that control how a service functions above and beyond the ability for two parties to communicate using voice or data.

Mobile companies have invested hugely in 3G licences. To obtain a profitable return on this investment there must be added value in terms of applications and services. It is unlikely that customers will pay the corresponding revenues just for voice. The suite of services above and beyond basic voice will include multimedia (content), mCommerce, information and entertainment, conversational services, and the integration of combinations of these into complete solutions. There is a huge opportunity in non-voice 3G terminals — everything from personal digital assistants, MP3 players, car-management systems and vending machines.

One of the desired concepts of the UMTS/3G network is the virtual home environment (VHE) [2]. VHE should allow a customer to access their services via different network interfaces, different terminals and while roaming. The concept of a personal service environment allows users to configure the properties and delivery of the service from a common, home-based, profile. Service control in 3G sets out to provide the flexibility and configurability to support the VHE concept.

The terms CS domain, PS domain and IM domain are used extensively in this chapter:

- CS domain is the circuit-switched signalling and switching network based on GSM;
- PS domain is the packet-switched data services introduced into GSM networks using a standardised overlay general packet radio service (GPRS) network;
- IM domain is the IP multimedia (IM) domain and includes GPRS enhancements for the support of real-time IP-based voice and data (multimedia) services.

Service control is one of the least developed aspects of the Release 4/5 architecture. Therefore, while this chapter builds upon the network architecture that is being developed in the 3GPP standards, the majority of the service control concepts discussed here represent proposals that have yet to be agreed in the standards. These concepts draw upon supplier directions and developments in bodies other than 3GPP (e.g. IETF [3]) — hence there is some justification for them being proposed.

7.2 Background

Before the future of service control can be defined it is useful to review the current technology for providing service control to see what lessons can be learnt. It is also essential to consider what the future mobile network will look like to know which resources and components the service control will be acting upon. Most importantly there needs to be a clear idea of what services are required and the level of demand to be met.

7.2.1 Appraisal of Intelligent Networks

Mobile networks have built upon the work of intelligent networks (IN) [4] from the fixed networks and developed their own variation called CAMEL [5]. CAMEL provides a means to link into the set-up or management of calls in the mobile switching centre (MSC) and control the operation of the call. The GSM CAMEL feature has been developed so that operator-specific services can be supported for customers who roam to foreign GSM networks.

IN in the fixed world is mature and is currently responsible for delivering a range of profitable services; but IN has never really lived up to its promise of providing an efficient, rapid mechanism for reusable services. Many services now exist, indeed in BT's UK fixed network there are many IN platforms providing a complete raft of services. IN, in general, has been complex to implement. This is in part due to the disparity in PSTN implementations, which has meant that, while typically the majority of an IN service is standard, some part always needs some form of specialist development to the switch or interface and this often delays service roll-out and increases the cost. Mobile networks are more standardised and hence should not suffer from the same problems.

IN distributes previously switch-embedded functions across network components allowing open procurement for different components. Thus service execution and interactive voice response (IVR) platforms may be bought separately from switch manufacturers and from suppliers that are more computing oriented. However, this distribution has introduced complexity due to interworking between switch features and the IN, and a distributed service design which necessitates expensive non-circuit-related signalling networks.

The technology that is used in IN has attempted to be more IT in nature, but has always been particular to the telecommunications industry. Consequently, while the concepts are similar, the telecommunications world has developed its own niche solutions, making them expensive.

In summary IN/CAMEL solutions have been a success in providing a means to realise services in 2G and fixed networks. However, while the concepts of open control interfaces and interaction with call set-up are reusable in 3G networks, service control needs to be less specialised in the technologies it uses so that there is an improvement upon IN.

7.2.2 Future Network Architecture

The focus for this chapter is on the 3G Release 4/5 (from the 3GPP standards) networks that are expected to be deployed circa 2004. Chapter 6 describes the IM domain for voice in a Release 4/5 network. This network will be a VoIP network with managed quality of service, using the GPRS nodes (SGSN, GGSN) to provide mobile IP access to an IP core.

The VoIP network, as standardised in 3GPP, will use the session initiation protocol (SIP) [6, 7]. SIP is an open Internet protocol for session initiation, following the model of HTTP (similar model and message structure). It is enabled to provide mobility, and aims to deal with the signalling associated with establishing sessions between end devices. Once a session is established then many different types of media exchange may be used.

3G operators will continue to offer GSM (2G) solutions for a considerable time, partly because of the investment that has been put into 2G and 2.5G (includes WAP and GPRS), and because 2G will still be required to provide complete geographical coverage.

New radio technologies in the UTRAN (GSM EDGE and UMTS WCDMA) mean that from a radio perspective roaming is becoming more complex. New core network technologies, such as PS and IM, mean that, from a service control perspective, roaming is also becoming more complex. Continuing the success of roaming achieved in GSM is a significant challenge and will be dependent on continued effort by operators and manufacturers in creating the necessary high-quality standards for UMTS.

The main emphasis of this chapter is on mobile networks but there are still many benefits that accrue both to the customer and to a combined fixed/mobile network operator, such as BT, from attempting to share common infrastructure components between networks. Seamless services have been much discussed but the concept runs into problems due to detailed network disparity, different network governance, and product portfolios. While not necessarily attempting to use the same network components between fixed and mobile networks, it is clear that there are many similarities in the technologies adopted by the fixed network.

As forecast by Ovum [8], by 2005 a quarter of the calls that would have been made on fixed lines (today) will instead be made by mobile. Telephony is currently the most important real-time service and is the most demanding of service control technologies. Hence, with telephony moving to mobile networks, the demand for service control is greater in mobile networks.

7.2.3 What Services?

Service control is required to provide services — so what are the real-time services required in 2004?

Within BT a number of customer scenarios have been identified (e.g. for the youth and SME markets). These attempt to provide a description of how the future 3G network will be used in terms of representative individuals for different market sectors. To fulfil these customer requirements, three service categories can be identified which contribute to delivering the scenarios:

- inherent from the network (connectivity, addressing, mobility, location, supplementary services);
- applications (distinction here is for non-real time, e.g. browsing, games, eCommerce);
- real-time services (pre-pay, call/session re-routing, call queuing, conferences).

It is the provision of the real-time services and the integration of these with the applications and network services where service control has a role.

The objective is not to predict all the individual services that will be required, and directly design the network to deliver those identified. Instead the emphasis is on establishing the type of resources, data models, control interfaces and performance that will be required from the service infrastructure.

While the services introduced above are oriented primarily towards the customer, there are other network capabilities for which the network provider will use service control. These include routing optimisation, internetwork call accounting and operator/customer service.

7.2.4 Wholesale at Every Level

Given the huge investment in 3G licences, there is some debate as to whether operators should open their networks to third party service providers or not. Keeping them closed means that operators singularly control the delivery of services and content — but will this approach allow customer demand to be satisfied and will it ensure sufficient network usage? It is unlikely that a single operator will be able to provide the transmission and radio infrastructure as well as all the rich content (and all pieces in the middle). Using a combination of owned services, partnered services and third party services will allow all customer requirements to be satisfied — from the customers you own, through the technologies you understand, to the customers you did not know existed or market sectors and technologies with which you are not familiar.

IP networks are more open since the protocols available at the access are the same as used in the 'core', and, so long as physical access to the IP is available, a service provider is at far less of a disadvantage compared to the network provider. However, if the network operator supports the service provider, it can provide more of the infrastructure that represents value-add than can be charged.

For network operators to satisfy the wide range of customer demands it is preferable that they do not take a 'walled garden' approach to their networks — better to embrace the access to third party service providers to deliver the complete range of services. It is wise to construct the 3G network such that wholesaling (access by third parties) is available at every level, e.g. access to fibre, access to call control, access to applications or provision of content. This will ensure efficient utilisation of the network infrastructure. Service control is included as one of these levels.

There are specific technologies that allow access to be provided to service control components, in a secure and managed way. These include the network API, Parlay (the variant specified by 3GPP), Open Service Access (OSA) and the Java variant, JAIN. These technologies are discussed in more detail in Chapter 8.

Providing wholesale access as an afterthought is difficult. It needs to be considered when a network is designed and implemented — in this way the same interface can be used for retail as well as wholesale, capacity expansion is easier, and interfaces can be made sufficiently secure and appropriate for third party use.

7.3 Intelligence Strategy

Over the past year an intelligence strategy has been put together within BT that defines how network operators can use their assets in IN, how they can adopt new service techniques given the changing nature of networks. The following subsections outline the strategy.

7.3.1 Evolution of IN

In the 3G network there will still be services offered today by IN platforms that will need to be continued. Customers will expect to see some services continuing to be offered by both the CS and IM domains. Hence services will naturally need to span the CS and IM domain. A good example is pre-pay — customers will not expect to have different credit accounts to deal with roaming between CS and IM domains, particularly when they have no control over where the domain coverage exists. Pre-pay is currently being implemented in many mobile networks via CAMEL, since it provides one of the most appropriate means of realising a secure implementation of pre-pay.

The evolution stage needs to be treated with great care. Any services evolved into the future network should be carefully selected, as there is a need to limit 'baggage' while providing the new service packages that are expected from UMTS.

7.3.2 IP Service Control

The clear focus of the strategy is to take advantage of IP as the underlying network technology and to use IP-oriented technologies to provide service control. In particular, this means looking at the way services are implemented using SIP. One of the most important attributes of VoIP is the separation of bearer control (media) from signalling control. This avoids one of the main restrictions of Signalling System No 7 (SS7) networks such as the PSTN. In CS networks, service nodes are a self-contained means of providing service control, but they suffer from the major cost penalty of having to trombone the voice circuits for the duration of the call, even if they only want to be involved in the call set-up phase. Many SIP service-control platforms can be involved in the establishment of the session with the eventual media stream only needing to be connected between the end terminals (or gateways). This optimal use of media streams and the simplicity of SIP signalling means that service platforms can be cascaded together with little cost or performance penalty for the duration of the call.

7.3.3 Service Control is not Just About Voice

Service control provides flexible mechanisms to create and execute control of real-time services. Packet-based network elements now have interfaces that allow for control of them in real time — to control admission, quality of service and packet priority. In general, this capability can be termed policy control and can be implemented via protocols such as the common open policy service (COPS) [9]. Hence this part of the strategy is to combine the service control capabilities and apply them to the policy control interfaces, i.e. data intelligence.

7.3.4 Choice of Technologies

To enable reductions in the cost and wider access to a larger development community, it is preferable to buy service solutions not from specialist industries such as telecommunications suppliers but instead from the much larger industry. To allow this to happen the technologies that are used to build the 3G network should be more compliant with the software industry. A simple test of this is to consider how many developers exist that can develop applications for SS7 interfaces compared to the number that can develop applications in Java and HTML. Given the move to an IP infrastructure, it is possible to use more software-industry-compliant technologies.

In the IM domain, the adoption of IP as a transport implies that there is no need for the specialist hardware (E1 interfaces) currently used for switches and IN platforms.

While there is still a very demanding requirement for high performance and availability, there is an opportunity to implement call servers and application servers on standard computing platforms and data centres or 'telecoms hotels' (see Young et al [10]). This has three potential implications for the 3G network:

- the adoption of a common platform to host telecommunications and Internet applications forces the two communities to share the same infrastructure, thus easing integration;
- the data centres provide network management (NM), provisioning and security facilities which will avoid the replication of separate telecommunications management systems — the current experience of implementing IN is that OSS and NM represent a significant cost, whereas operating from a common applications platform will allow these costs to be shared between the (data centre) users;
- many operators have data centres across the globe — by hosting a network component in a foreign country, it provides a mechanism to provide 'home' servers that are physically more local to visited locations, thus helping to reduce the cost and delay of returning signalling to a home environment.

7.3.5 Terminal Intelligence

Mobile terminals are increasing in intelligence and computing facilities. Already, via embedded microcode, the SIM tool-kit, WAP and MExE (mobile station application execution environment) services can already run directly on the terminal. The service provider has the opportunity to implement services either via service control in the network or in the terminal. The real opportunity is to use the two together, leveraging the individual strengths.

7.4 3GPP Service Control

The 3GPP Release 4/5 architecture, which delivers the all-IP network, identifies the need to link service control into the call control and GPRS components, but it has been the least well defined part of the architecture. This section describes some of the ideas being pursued in 3GPP to standardise the service control interfaces.

7.4.1 Service Control in the CS and PS Domains

CAMEL is the main standardised capability in the core network for control of operator-specific services when roaming. It is based on IN capability sets (CS-1/CS-2). CAMEL has been developed in phases, each phase adding new capabilities while retaining backward compatibility with earlier phases.

Figure 7.1 illustrates the CAMEL Phase 3 network architecture. CAMEL Phase 3 supports circuit-switched voice and data services at the gateway MSC (GMSC) (mobile terminated calls and unconditional forwarded calls) and visited MSC

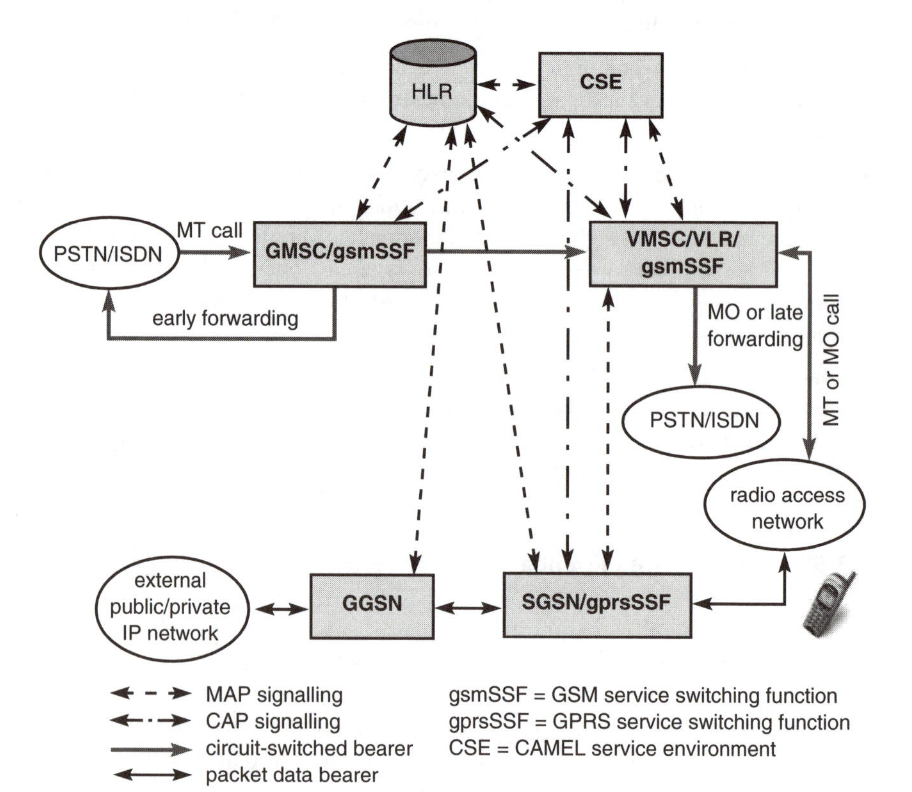

Fig 7.1 CAMEL Phase 3 network architecture.

(VMSC) (mobile originated calls and conditional forwarded calls). Many different services can be controlled using CAMEL. Some of the most common services include location services, pre-pay, mobile VPN, and short code number translation services. CAMEL also includes access to a specialised resource function (gsmSRF) which provides user interaction mechanisms such as announcements and voice recognition (not shown in Fig 7.1).

Services in the PS domain are controlled through interworking with GPRS at the SGSN (gprsSSF). Services that can be supported in the PS domain include pre-pay, barring, and location services. The server has control of GPRS PDP contexts and GPRS attach/detach at the SGSN using CAMEL capabilities. Access for third party service control is supported using a CSE-based API such as OSA.

7.4.2 IM Domain Network Architecture

Figure 7.2 focuses on the architecture of the IM domain, illustrating key network functional entities and interfaces. This diagram reflects current thinking and so may not represent the final design.

The architecture in Fig 7.2 is depicted from the perspective of a mobile-initiated multimedia session. SIP signalling is used from the terminal through to the S-CSCF (serving call state control function) via the SGSN (serving GPRS support node) and

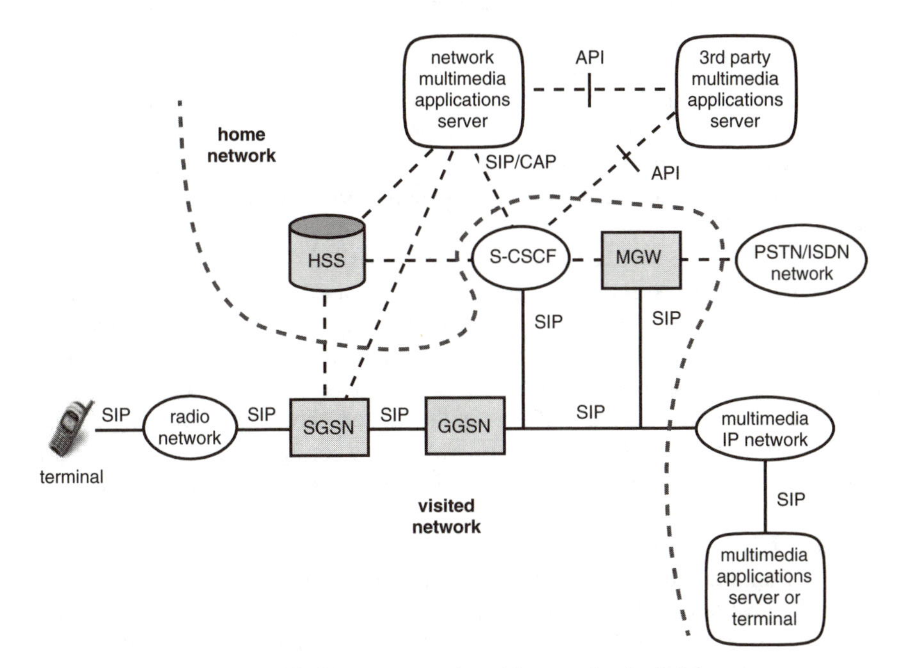

Fig 7.2 Simplified 3GPP network architecture for the IM domain.

GGSN (gateway GPRS support node). The GSNs are enhanced GPRS network entities based on earlier GSM specifications. The CSCF provides the call-control, resource-control and registration services. The home subscriber server (HSS) is a database that can be accessed by the SGSN, GGSN, S-CSCF and multimedia applications server (MAS). The HSS may include HLR functionality so that IM domain customers may roam to GSM networks that do not support the IM domain. The terminal must register with a CSCF known as the serving CSCF (S-CSCF), before network or third-party applications can be accessed. The HSS will be used to store user profiles in addition to the location and registration data. The S-CSCF may be part of the home network or visited network. If it is part of the home network then a CSCF incorporating a proxy server is used in the visited network to route SIP messages to the S-CSCF. The proxy CSCF is a SIP proxy server; however, the proxy CSCF must perform other functions not normally associated with a SIP proxy. The proxy CSCF must handle emergency calls, charging and resource allocation, such as access to media gateways. The proxy CSCF does not have any knowledge of the customer's identity.

7.4.3 Service Control in the IM Domain

Within the 3G Release 4/5 network the three most significant components from a service control point of view are the terminal, the CSCF and the HSS. A user will want to access services from their home network and take advantage of any local services from the visited network. A network operator wants to offer services to its own customers, but there is also an opportunity to offer services to roaming customers. The combinations of how service control can be accessed from serving and proxy CSCFs is shown in Fig 7.3.

In Fig 7.3(a), the serving CSCF is in the home network and calls are triggered to the home MAS.

In Fig 7.3(b), the user has elected to use the visited CSCF as the serving CSCF but actually receives services from the home-based MAS. This intra-country network service control interface is currently being realised in 2G networks using CAMEL. This configuration would allow a user to dial the normal simple code to pick up voice messages, e.g. 901.

In Fig 7.3(c), the serving CSCF is in the visited network but the user elects to receive service from the local operator's MAS. This configuration could be used, for example, to bill all calls to a temporary credit account being used in the visited network.

In Fig 7.3(d), the serving CSCF is in the home network but the user is accessing services from the proxy CSCF. The proxy has no knowledge of the user, hence they cannot be profile based (although some form of profile may be received from the terminal — see section 7.4.7). However, there are many services that could be provided. For example, the simple service of 'find local restaurants' could be provided from the proxy — there would be little point receiving this service from the home network. In addition, this figure shows service control also being applied to the serving CSCF, and, as with all these configurations, all combinations are possible.

In Fig 7.3(e), the serving CSCF is in the home network but the user receives services from a local MAS. The user is likely to be in the same network as the MAS but it could be that network/service providers make their services available, outside their network, irrespective of where the user is located.

As each of the options (a) to (e) in Fig 7.3 can apply to the caller or calling party, there are actually many combinations when the end-to-end call is considered, e.g. the calling party may be using configuration 7.3(a) while the called party is using

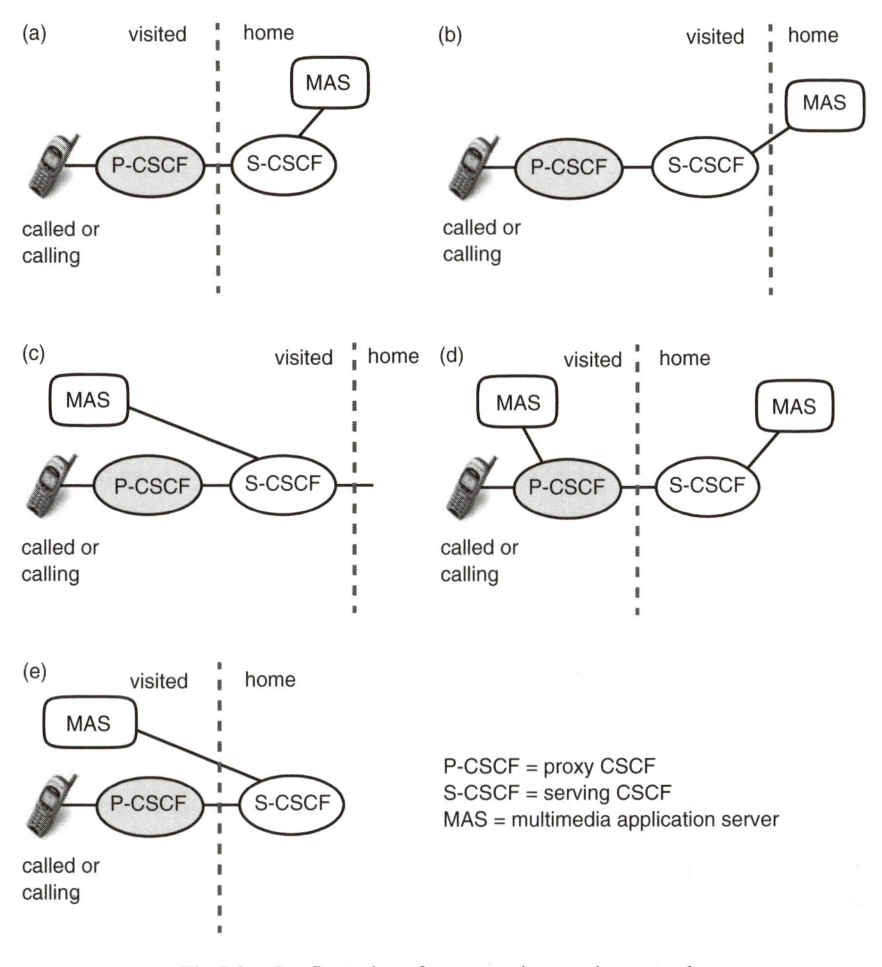

Fig 7.3 Configurations for connecting service control.

7.3(b). These configurations make no distinction whether the service is being provided by the network operator or by a third party (via a network operator).

7.4.4 Service control interfaces

The interfaces that need to be defined for service control are shown in Fig 7.4.

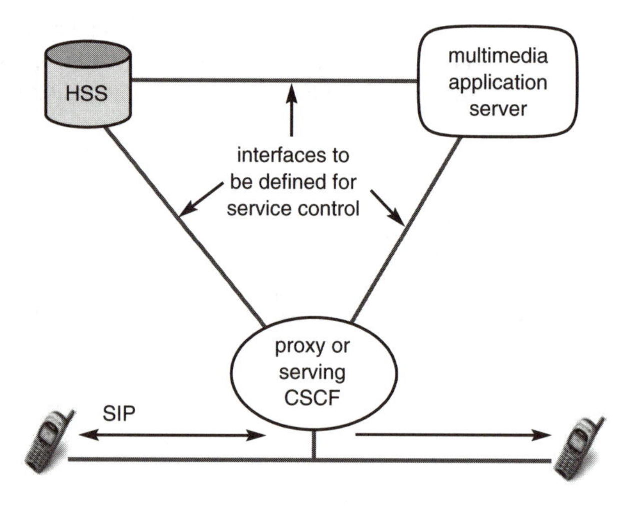

Fig 7.4 Key service control interfaces.

Given the requirements for the evolution of CAMEL, support for third party interfaces and the desire to exploit IP technology, the approach has been to adopt a flexible approach on the interfaces available and to support three different types of service control interface at the CSCF:

- SIP;
- CAMEL (although using IP as the signalling transport);
- application programmable interfaces (APIs).

The implementation of these interfaces is not mandatory.

In order to support these service-control interfaces some form of model of the CSCF is required that determines how these interfaces interact with the session set-up. A simple model has been proposed for the CSCF and is shown in Fig 7.5. This is not meant to design or standardise the implementation of the CSCF but does determine the data necessary for the operation of the service control interfaces. Here the IN concept of triggering can be seen being applied. There must be a method of defining the criteria for interrupting session set-up while another function is referred to determine how future aspects of the session are achieved.

Illustrated in Fig 7.5 are the proposed S-CSCF internal functions and external interfaces. The session control function decides how a SIP invite is to be handled, i.e. it determines which service control will be used — CAMEL, API or SIP server. The decision is based on information contained in the SIP invite message, such as caller identity, service data and customer profile data. Customer profile data is downloaded from the HSS to a local serving profile database (SPD) when the customer registers with the CSCF. The SPD is effectively a cached version of the data from the HSS.

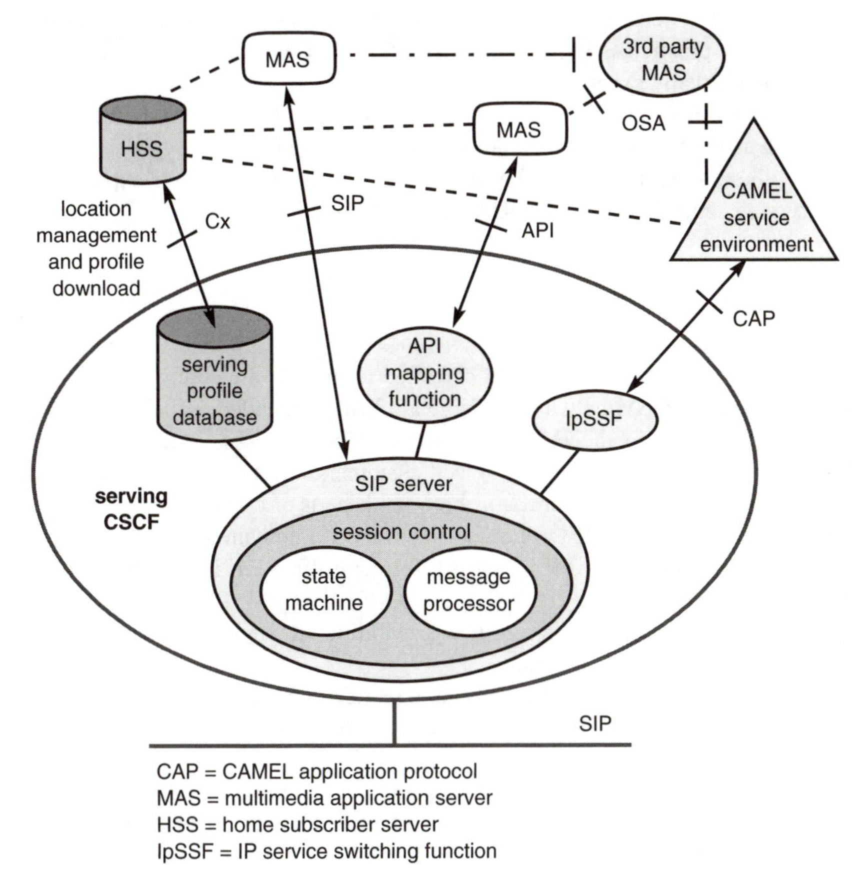

Fig 7.5 CSCF model for service control.

It is very likely that a given call may want to have multiple interactions, i.e. triggers with the service control, e.g. triggering to the CSE over CAP to do pre-pay, followed by triggering to the MAS over SIP to perform conditional-based routing. Therefore, on receiving a response from one trigger, the resulting invite must refer to the session control to see where the session is next sent.

The IpSSF provides the interworking between the interface to the SIP server and the CAMEL protocol. The IpSSF exhibits a call state model to the CSE, e.g. allowing dynamic event-detection points to be enabled and generally support the CAMEL operation set.

The mapping function would encapsulate the SIP server and present it to the MAS in whatever API format is required — this would include dealing with the distributed computing environment being used, e.g. CORBA or remote method invocation (RMI).

For the SIP interface, no interworking or translation function is considered necessary. In practice, some SIP services (e.g. call-processing language (CPL) scripts) may actually execute inside the CSCF.

Each of the MASs may expose their service capabilities to third parties using APIs such as OSA. The MAS is therefore acting as a gateway.

The following sections describe in more detail the intentions for exploiting the three service control interfaces — CAMEL, APIs and SIP.

7.4.5 CAMEL in 3G

This CS domain capability has been extended to include the control of services in the PS domain for GSM networks. It is being further extended to include control of multimedia services in the IM domain for UMTS networks. Extending CAMEL allows network operators to provide customers with services that behave consistently when the customer roams between islands of UMTS and GSM.

Developments for CAMEL interworking in the IM domain are currently focused mainly on IP telephony. CAMEL Phase 3, for controlling services in the CS and PS domains, is currently being enhanced to CAMEL Phase 4. Phase 4 includes new capabilities for the CS domain and new capabilities to support voice over IP and multimedia calls in the IM domain.

Service control is achieved by introducing a new interface between the CSE and the S-CSCF based on the CAMEL protocol. This work is at an early stage of development in 3GPP.

One of the challenges for CAMEL in the IM domain is that, during the life of a session, extra media interactions can be added into the session. Although CAMEL is well able to control telephony communication, it is, however, less able to control, for example, a chat session.

There are multiple possible solutions to this problem. The CSCF may block any additional non-telephony service requests if CAMEL is already active in a session. Alternatively the IpSSF could monitor subsequent SIP transactions and, if it detects a media exchange which it does not understand, it would drop out of the session in a clean way.

7.4.6 Network APIs in 3G

Application programming interfaces (APIs), such as Parlay, JAIN (Java integrated network) or Open Service Access (OSA), provide an interface which is readily integrated into application software that allows the application to use the services of a network without being concerned with the exact nature of the network.

These APIs have capabilities via their framework elements to allow the API to be accessed by third parties. These capabilities include authentication, discovery and security. An operator can choose either to use an API for their own applications or to make the interface available to third parties. If it is to be made available to third parties, it is advisable that this is done via some gateway function that will allow the framework capabilities to be suitably implemented.

The inclusion of an API interface into the CSCF also reflects a supplier trend in application server technology. Suppliers want to encourage application writers to use their platforms and they also want to reuse the application server with multiple networks, i.e. SIP, H.323, PSTN. This portability capability is provided to allow for the control of multiple network interfaces. By adopting middleware and API architectures, they can provide programmability platforms that allow the rapid adaptability to different network types and allow the application writer to be divorced from the specific network technology. Offering an API interface on the CSCF allows for easy integration into programmability platforms.

APIs are already being offered on top of SIP stacks. As the next section describes, services can be written directly to SIP interfaces but some of these interfaces appear very raw and low level. Encapsulating the SIP interfaces with OSA can provide value by providing abstracted building blocks for putting services together.

7.4.7 SIP services

The most attractive of all the service control interfaces is the use of SIP. This is due to the wide-scale adoption of SIP and the simplicity and ubiquity of the technology it uses to realise services.

One of the real appeals of the SIP service technologies is that they can be integrated into WWW technologies that are used widely today. Through the adoption of these technologies the 3G network will be able to provide rapid and inexpensive services and, probably most importantly, there will be a mass of people with the ability to provide the wealth of services desired. In effect, the 'dot.com' creative spirit can be applied to the UMTS network.

There are at least five technologies for providing service control that are available when using SIP. Any combination of these technologies can be used together. As these technologies are defined as extensions to the basic SIP protocol, there is no guarantee that they will be provided by a network operator. Even if it is agreed in 3GPP to include them, it is likely that it will not be mandatory to support them. All the SIP technologies under consideration [6, 7] are shown in Fig 7.6.

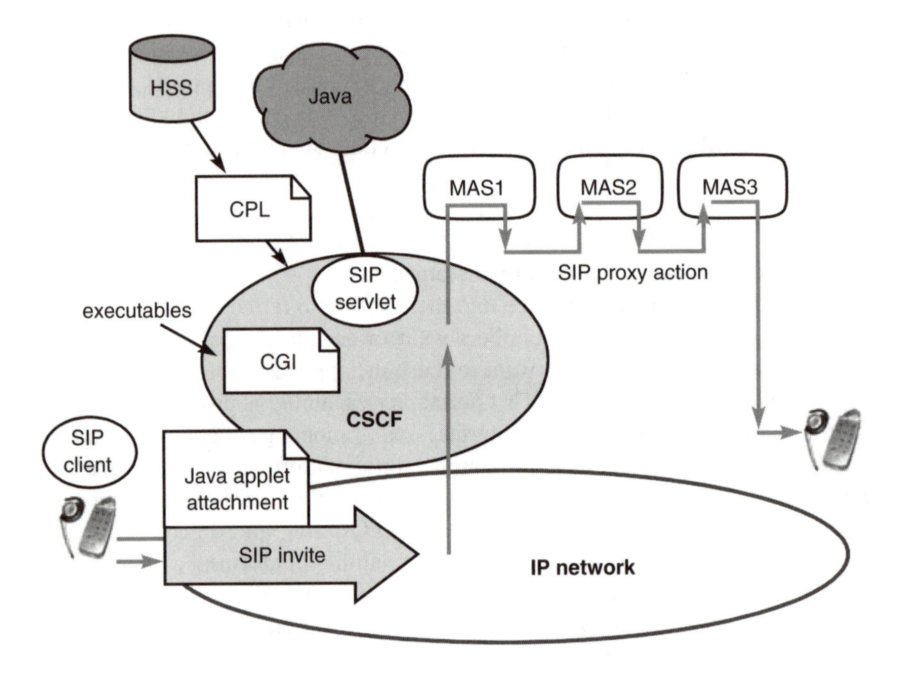

Fig 7.6 The SIP service opportunities.

7.4.7.1 SIP CGI

In the WWW, the common gateway interface (CGI) has been very successful in allowing dynamic applications to be linked into the otherwise static content-only information of a Web page. Using CGI the WWW is able to react to the user-initiated events and trigger eCommerce payments, reservations, telephony calls, etc. Due to the similarity of HTML and SIP it has been a natural extension to apply CGI to SIP.

A SIP CGI script has a number of key attributes. It can expose the content of all headers in a SIP request (this is essential, because, within SIP, the header is of great importance, carrying critical information about the session, including caller, callee, subject, registration parameters, call routes, etc). It can also create all parts of a response, including headers and message bodies.

A single CGI script may execute a number of times during a transaction and, by use of CGI triggers, can specify at which point it will have control for the remainder of the transaction. These triggers can include timers and triggers within the headers of the SIP request. Alternatively the script can be set to execute on every message.

The actual script could be any executable produced by languages such as C and Visual Basic or script languages such as PERL. The similarity with HTML CGI means that existing developers will be readily able to develop SIP CGI scripts.

Upon receiving a SIP request the server will pass the body of the message to the script and set up environment variables containing the information on the message headers. The script will then generate some output data which is then processed by the server to create a response message if required.

7.4.7.2 SIP Call-Processing Language

Call-processing language is a very simple XML-based language that can be used to describe and control Internet telephony services. It is actually protocol independent (i.e. it could also be used with H.323), but it fits particularly well with SIP. The aim of CPL is to provide a secure and straightforward method of providing simple call services that can be easily created by graphical tools, e.g. by tabular WWW entry. In the context of 3G service control this provides an appropriate a way of implementing some of the personal profile concepts associated with VHE.

A CPL script is built around call-processing actions. A call-processing action is a structured tree that describes the decisions and actions which the server performs on a call event. There are two types of call-processing action — top-level actions, which are triggered by signalling events, and sub-actions, which can be called from other actions and are grouped into definitions and references. Sub-actions are effectively like sub-procedures in normal computing languages.

Top-level actions can be divided into incoming (an action when the call destination is the script owner) and outgoing (an action when the call originator is the script owner). The CPL script could be downloaded into the serving CSCF from the HSS when a user registers. The four categories of CPL nodes are switches, location modifiers, signalling actions and non-signalling actions.

Table 7.1 summarises the list of node types available and demonstrates the compactness, yet power, of the language.

CPL is designed to be easily created and edited by graphical tools and, as it is XML based, it is easily parsed. It is considered safe as a scripting language as it does not permit loops, just conditional statements.

It is worth noting the power and simplicity offered by CPL compared with embedded switch control and IN concepts in current 2G networks. For example, it is not inconceivable that a network provider could allow third parties to provide CPL scripts to be loaded into the S-CSCF. The safety of CPL combined with the fact that it can be verified without execution makes this a plausible possibility. A service provider would then need only a portal to create new customers and a Web page to allow users to create, modify and delete their call-handling services. The service provider simply downloads the CPL script to the HSS (for onward distribution to the CSCF). The addition of the OSA framework could provide a means to provide authentication, security and discovery. By comparison, in 2G networks a service provider has no choice but to develop real-time applications or acquire a complete service node to offer similar functionality.

Table 7.1 CPL node types and function.

CPL node type	CPL node	Description
Switches	Conditional branching based on information from the original request or from the call.	
	Address switch	Based on one of the addresses in the call request.
	String switches	Based on free-format text in the call request.
	Time	Based on time and date. Includes extensive sub-parameters.
	Priority	Based on the priorities specified in SIP, e.g. emergency, urgent.
	Otherwise	Special: If none of the other cases match.
	Not present	Special: If switch variable not present in the request.
Location modifiers	The set (could be multiple-forked destinations) of locations to which a call should be directed.	
	Explicit	Specified locations to add to the location set.
	Look-ups	Refers to locations defined in an external source.
	Filters	Remove locations from the set.
Signalling actions		
	Proxy	Forward the call to the current sent of locations. The result of the forward can have further cases specified of: time-out, busy, no-answer, redirection, failure, recursion, reordering.
	Reject	Reject the call attempt.
	Redirect	Redirect the call attempt to the current set of locations.
Non-signalling actions		
	Mail	Mail the status of the CPL script to the user.
	Log	Log information about the call to some form of storage.

7.4.7.3 SIP Servlets

SIP servlets are effectively a Java extension API for SIP servers. It is a portion of Java code that interacts with a SIP server to control or influence session processing. Incoming requests and responses are associated with the servlets. The CSCF communicates to the servlet (running on the MAS) by passing objects representing SIP messages. The servlet has access to all parts of the SIP message (header and body as well as request and status line) via the objects. The servlet can then answer, proxy requests (to multiple destinations if required), create responses, forward incoming requests or handle them directly. Servlets can also initiate new SIP transactions of their own.

The CSCF will effectively pass an object representing the request to a servlet. The servlet then decides whether to respond to it, proxy it, or let the CSCF perform its default action.

While servlets provide a suitable abstraction of SIP it may be necessary to further encapsulate them into a network API, perhaps JAIN given the Java environment, to make it easier for the application writer.

7.4.7.4 Java-Enhanced SIP

Another possible solution is the Java-enhanced SIP (JES). SIP messages are extended to carry Java applets (plus their state and run-time contexts if needed). These Java mobile objects are run before the CSCF takes any other action. A 'require' message is used within all JES SIP messages to indicate that JES must be supported to process the message. The Java applet or Java mobile agent is stored either in a multipart MIME section of the message body or at a URL pointed to within the message.

The receiving CSCF retrieves the mobile agent within a SIP message and it must immediately extract it and execute it on a virtual machine before processing the SIP message any further (or if the content is a URL, it must retrieve the agent from the URL). The agent can then interact with the proxy and instruct the proxy to carry out commands.

A Java SIP API provides an interface into a SIP proxy to allow applets and agents access to SIP messaging functions on the proxy. This is a similar approach to SIP servlets.

The use of JES in 3GPP is yet to be agreed. Like CPL it provides a mechanism for users to apply their profile. However, given that the applets are executable, there is more power to support a greater range of services, but at the same time more challenges in terms of security.

7.4.7.5 SIP Proxy Capability

One of the fundamental mechanisms of SIP is to proxy on a request to another proxy server where further processing can take place. Hence a very simple but powerful way to provide service control is for a CSCF to forward the SIP invite to the MAS. The MAS can then provide some form of service, which again could be CAMEL, APIs or one of the SIP service technologies, and ultimately it too can further forward the message on to another MAS.

There are two basic configurations for proxying MAS in relation to the CSCF.

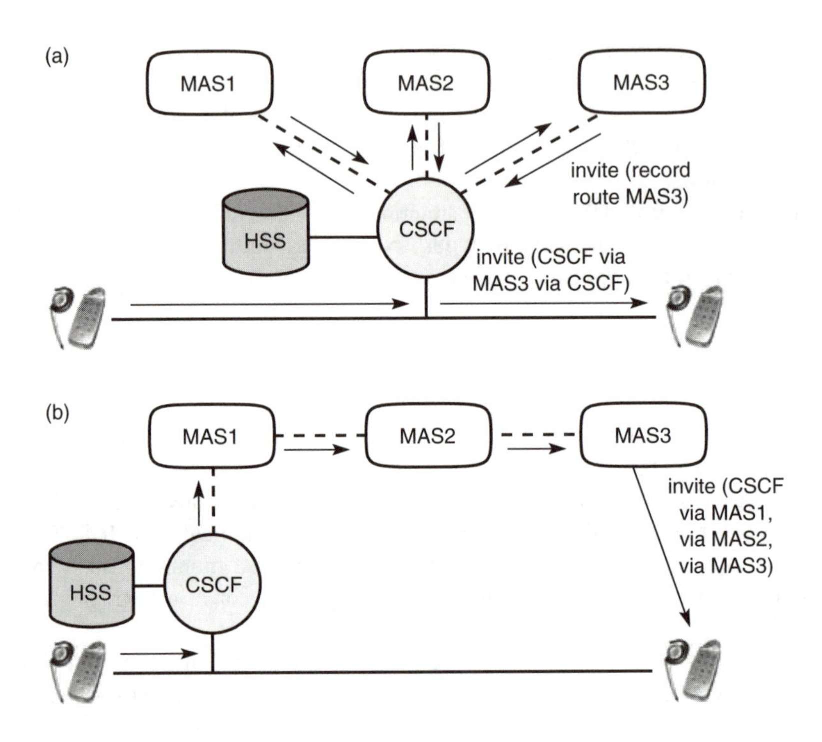

Fig 7.7 MAS proxy configurations.

In Fig 7.7(a) the CSCF proxies invites in turn to a series of MASs. This allows the sequence of services (resident on the MASs) to be determined from the user profile held in the HSS. In practice, it may not be the CSCF but a separate distribution server which has access to the HSS. As some services may only want to be involved in modifying the location, for example MAS1, it requests the CSCF to redirect the invite to MAS2 such that MAS1 is dropped out of the list of locations and does not see future transactions. MAS3 may want to see future transactions, e.g. the termination of the session in a pre-pay service, and therefore adds its own location into the record route so that future transactions are returned to it.

In Fig 7.7(b) the CSCF forwards the invite to a series of MASs. The preceding MAS needs to know at least the next hop in the chain. In this configuration the sequence and list of services is not controlled from a single profile.

In practice, both of the configurations in Fig 7.7 can be used together, where the MAS locations shown in the (b) configuration are actually islands of (a). This would allow a user to receive services from different service environments provided by different operators.

7.5 Conclusions

Real-time service control has a major part to play in realising an integrated, useful and stimulating experience for the customer and as a secure, profitable, and flexible tool for the network and service provider. The strategy being followed in developing service control in the Release 4/5 network is encouraging flexibility to allow for the evolution of CAMEL, inclusion of programmable API platforms and using the power of SIP technologies.

The chapter has described service control architectures that demonstrate a user-centric view, allowing the services to be selected from a user profile and in collaboration with the user's terminal.

The richest service environments will be those having an open approach to providing services. Adopting standard interfaces that are attractive to the application developer community allows the operator to have access to a broad set of services. Additionally the interfaces should be open so that third party service providers can add to the services provided by the network operator, increasing the attractiveness of subscribing to a particular network and accessing a greater range of innovative services.

References

1 3GPP Web site — http://www.3gpp.org

2 3GPP: '*Virtual home environment*', TS 22.121 V3.2.0 (March 2000).

3 IETF Web site — http://www.ietf.org

4 Abernethy, T. W. and Munday, A. C.: '*Intelligent networks, standards and services*', in Dufour, I. G.: '*Network Intelligence*', Chapman & Hall (1997).

5 3GPP: 'CAMEL', TS23.078 (2000).

6 IETF: '*SIP: Session initiation protocol*', RFC 2543 (March 1999).

7 Wisely, D. R.: '*SIP and conversational Internet applications*', BT Technol J, **19**(2), pp 107-118 (April 2001).

8 Ovum Forecasts: '*Global Telecoms and IP Markets*', (January 2000) — http://www.intellact.nat.bt.com/intellact/index.htm

9 Boyle, K. et al: '*The COPS (Common Open Policy Service) protocol*', Internet Draft <draft-ietf-rap-cops-05.txt> (2000).

10 Young, S., Gardiner, D. and Walton, S.: '*Telecoms hotels, collocation and data centres — hosting the new e-economy*', Ovum (July 2000).

8

THE OSA API AND OTHER RELATED ISSUES

R M Stretch

8.1 Introduction

3GPP was conceived some two to three years ago with the express purpose of accelerating work on mobility. Existing standards organisations such as the ITU-T and ETSI had been successful in producing a vast array of international standards for different technologies; however, the speed with which this was attained was slow, especially when compared to bodies such as the IETF. It therefore remained a target of the mobile industry to produce coherent specifications in a time frame that was ahead of those within the standards industry — hence the formation of 3GPP. At the end of 1998 and the beginning of 1999 3GPP formed a group known as OSA, Open Service Access. The purpose of this group was to focus 3GPP's efforts on defining an architecture in support of the virtual home environment (VHE) — the need here being to provide mobile users with access to their service offerings irrespective of their position within or outside the home network environment.

8.2 Virtual Home Environment

Virtual home environment is defined as a concept for personal service environment (PSE) portability across network boundaries and between terminals. The concept of the VHE is such that users are consistently presented with the same personalised features, user interface customisation and services in whatever network and whatever terminal (within the capabilities of the terminal and the network), wherever the user may be located. The first specification was known as Release 99. This was seen as a historical separation between networks of the 20th Century and those of the 21st Century. Release 99 would therefore be a link between the two. For Release 99, CAMEL, MExE and SAT are, for example, considered the mechanisms supporting the VHE concept.

The personal service environment describes how the user wishes to manage and interact with their communications services. It is a combination of a list of ‘subscribed to’ services, service preferences and terminal interface preferences. PSE also encompasses the user management of multiple subscriptions, e.g. business and private, multiple terminal types and location preferences. The PSE is defined in terms of one or more user profiles. The user profiles consist of two kinds of information:

- interface-related information (user interface profile);
- service-related information (user services profile).

8.3 OSA

The OSA defines an architecture that enables operator and third party applications to make use of network functionality through an open standardised API (the OSA API). OSA provides the glue between applications and service capabilities provided by the network. In this way applications become independent of the underlying network technology.

The applications constitute the top level of the OSA. This level is connected to the service capability servers (SCSs) via the OSA API. The SCSs map the OSA API on to the underlying telecommunications-specific protocols (e.g. mobile application part (MAP) and the CAMEL application part (CAP), etc) and are therefore hiding the network complexity from the applications.

Applications can be network/server-centric applications or terminal-centric applications. Terminal-centric applications reside in the mobile station (MS).

Examples are the mobile execution environment (MExE) and SAT applications. Network/server centric applications are outside the core network and make use of service capability features offered through the OSA API. (Note that applications may belong to the network operator domain although running outside the core network. Outside the core network means that the applications are executed in application servers that are physically separated from the core network entities.)

8.4 Parlay Influences the Specification

Just previous to the formation of the OSA group, an industry consortium know as Parlay set about defining a network-independent API. This was to be used between a network provider and a service provider for the express purpose of accessing third-party service applications without the need for designing those applications for each independent network technology, such as mobile, fixed narrowband, fixed broadband, etc. This was achieved by introducing ‘middleware’. Middleware provides the separation between the two environments (networks and applications) and also any necessary mapping functionality between the two different communications environments, the API and, say, MAP.

It was therefore this type of concept that OSA decided to adopt within their architecture. The Parlay group were in the process of defining a second set of specifications for their third party access API and had decided to include mobility. As there was no point in redefining the wheel, OSA took as their starting point a number of Parlay-defined API methods, such as call control and framework, and set about producing requirements for mobility. One problem with this approach was that the mobility API methods may have deviated from those being produced within the Parlay group. As some members of the Parlay group were also active participants in the 3GPP group, unofficial interactions were set up between the two groups which resulted in reducing the differences between the two APIs.

8.5 OSA Explained

In order to implement unknown end-user services/applications today, a highly flexible open service architecture is required. The OSA is the architecture enabling applications to make use of network capabilities. The applications will access the network through the OSA API.

Network functionality offered to applications is defined as a set of service capability features (SCFs) in the OSA API, which are supported by different SCSs. These SCFs provide access to the network capabilities on which the application developers can rely when designing new applications (or enhancements/variants of already existing ones). The different features of the different SCSs can be combined as appropriate. The exact addressing (parameters, type and error values) of these features is described in stage 3 descriptions. These descriptions (defined using the OMG Interface Description Language™) are open and accessible to application developers, who can design services in any programming language, while the underlying core network functions use their specific protocols. The aim of OSA is to provide an extendible and scalable architecture that allows for the inclusion of new service capability features and SCSs in future releases of UMTS with a minimum impact on the applications using the OSA API.

The standardised OSA API is secure, it is independent of vendor-specific solutions and independent of programming languages, operating systems, etc, used in the service capabilities. Furthermore, the OSA API is independent of the location within the home environment where service capabilities are implemented and independent of supported server capabilities in the network. To make it possible for application developers to design new and innovative applications rapidly, an architecture with open interfaces is imperative. By using object-oriented techniques, like CORBA, it is possible to use different operating systems and programming languages in application servers and service-capability servers. The service-capability servers act as gateways between the network entities and the applications.

The OSA API is based on lower layers using mainstream information technology and protocols. The middleware (e.g. CORBA) and lower layer protocols (e.g. IP) should provide security mechanisms to encrypt data (e.g. IPSec).

8.6 Overview of Open Service Access

Open Service Access (Fig 8.1) consists of three parts, as described below.

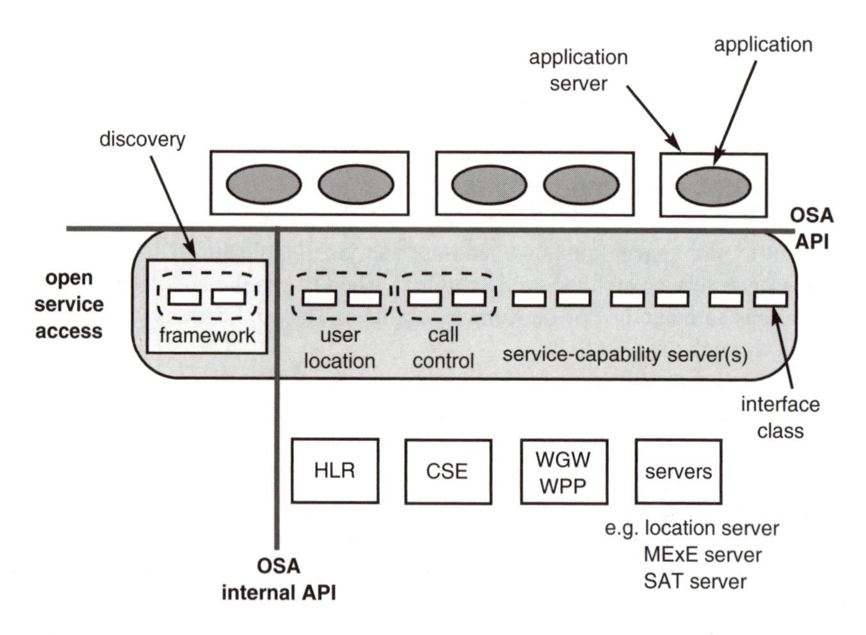

Fig 8.1 Overview of the Open Service Access architecture.

- Applications

  Applications, such as VPN, conferencing and location-based applications, are implemented in one or more application servers.

- Framework

  This provides the basic mechanisms that enable the applications to make use of the service capabilities in the network. Examples of framework service capability features are authentication and discovery. Before an application can use the network functionality made available through service-capability features, authentication between the application and framework is needed. After authentication, the discovery service-capability feature enables the application to find out which network service-capability features are provided by the service-capability servers. The network service capability features are accessed by the methods defined in the OSA interfaces.

- Service capability servers

  These provide the applications with service-capability features, which are abstractions from underlying network functionality. Examples of those features offered by the service-capability servers are call control and user location. Similar service-capability features may possibly be provided by more than one service-capability server. For example, call-control functionality might be provided by SCSs on top of CAMEL and MExE.

The OSA service-capability features are specified in terms of a number of interfaces and their methods. The interfaces are divided into two groups:

- framework interfaces;
- network interfaces.

The interfaces are further divided into methods. For example, the call manager interface might contain a method to create a call (which realises one of the service-capability features — 'initiate and create session'). It should be noted that the CAMEL service environment does not provide the service logic execution environment for applications using the OSA API, since these applications are executed in application servers.

The service-capability servers that provide the OSA interfaces are functional entities that can be distributed across one or more physical nodes. For example, the user-location interfaces and call-control interfaces might be implemented on a single physical entity or distributed across different physical entities. Furthermore, a service capability server can be implemented on the same physical node as a network functional entity or in a separate physical node. For example, call-control interfaces might be implemented on the same physical entity as the CAMEL protocol stack (i.e. in the CSE) or on a different physical entity — several options exist.

- Option 1

  The OSA interfaces are implemented in one or more physical entities, but separate from the physical network entities. Figure 8.2 shows the case where the OSA interfaces are implemented in one physical entity, called a 'gateway'. Figure 8.3 shows the case where the SCSs are distributed across several 'gateways'.

- Option 2

  The OSA interfaces are implemented in the same physical entities as the traditional network entities (e.g. HLR, CSE) (see Fig 8.4).

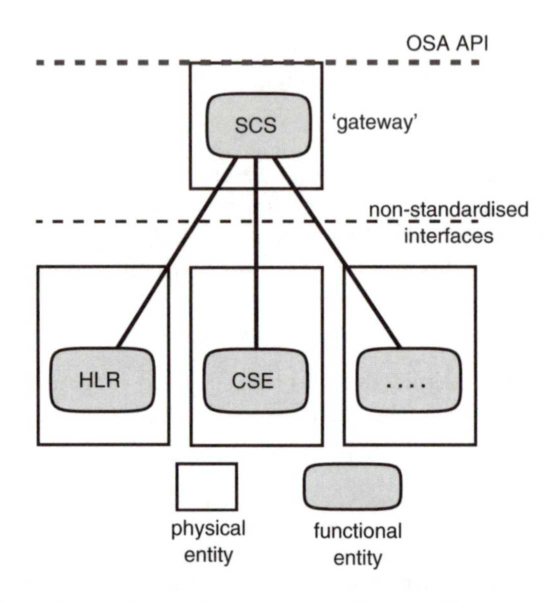

Fig 8.2 SCSs and network functional entities implemented in separate physical entities.

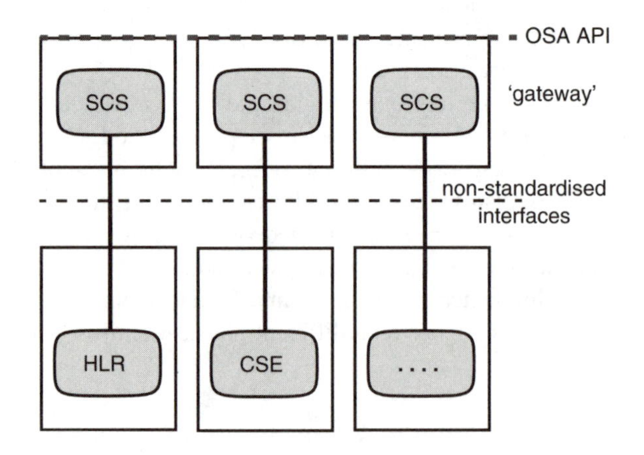

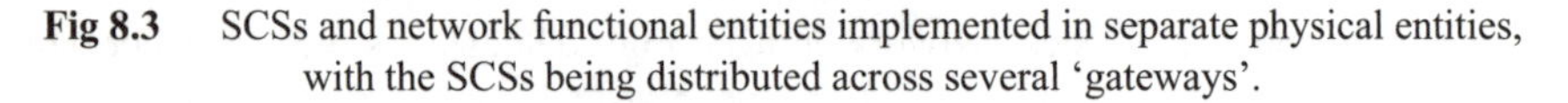

Fig 8.3 SCSs and network functional entities implemented in separate physical entities, with the SCSs being distributed across several 'gateways'.

- Option 3

  Option 3 is the combination of option 1 and option 2, i.e. a hybrid solution (see Fig 8.5).

It should be noted that in all cases there is only one framework. This framework may reside within one of the physical entities containing an SCS or in a separate physical entity. From the application point of view, it makes no difference which implementation option is chosen, i.e. in all cases the same network functionality is perceived by the application. The applications are always provided with the same set of interfaces and a common access to framework and service capability feature interfaces. It is the framework that will provide the applications with an overview of available service capability features and how to make use of them.

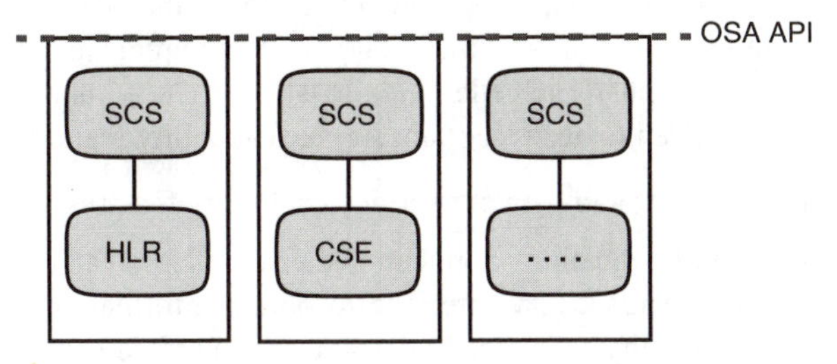

Fig 8.4 SCSs and network functional entities implemented in the same physical entities.

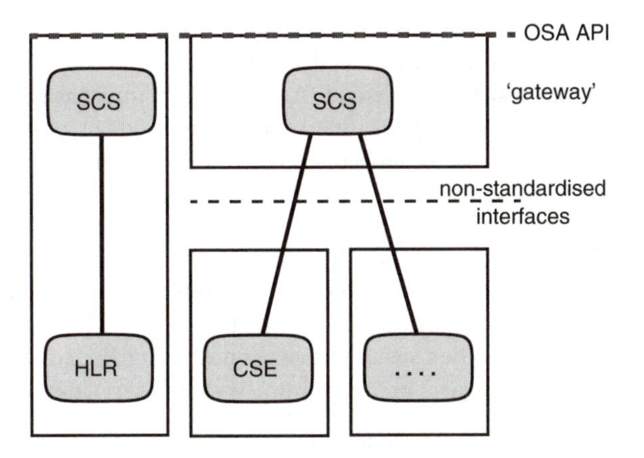

Fig 8.5 Hybrid implementation (combination of options 1 and 2).

8.7 Basic Mechanisms in Open Service Access

This section explains which basic mechanisms are executed in OSA prior to offering and activating applications.

Some of the mechanisms are applied only once (e.g. establishment of service agreement), while others are applied each time a user subscription is made to an application (e.g. enabling the call attempt event for a new user). The following are the basic mechanisms between application and framework.

- Authentication

  Once an off-line service agreement exists, the application can access the authentication interface. The authentication model of OSA is a peer-to-peer model. The application must authenticate the framework and vice versa. The application must be authenticated before it is allowed to use any other OSA interface.

- Authorisation

  Authorisation is distinguished from authentication in that authorisation is the action of determining what a previously authenticated application is allowed to do. Authentication must precede authorisation. Once authenticated, an application is authorised to access certain service capability features.

- Discovery of framework and network service-capability features

  After successful authentication, applications can obtain available framework interfaces and use the discovery interface to obtain information on authorised network service-capability features. The discovery interface can be used at any time after successful authentication.

- Establishment of service agreement

  Before any application can interact with a network service-capability feature, a service agreement must be established. A service agreement may consist of an off-line (e.g. by physically exchanging documents) and an on-line part. The application has to sign the on-line part of the service agreement before it is allowed to access any network service-capability feature.

- Access to network service-capability features

  The framework must provide access control functions to authorise the access to service-capability features or service data for any API method from an application, with the specified security level, context, domain, etc. The following describes the basic mechanism between framework and service capability server.

- Registering of network service-capability features

  SCFs offered by a service-capability server can be registered at the framework. In this way the framework can inform the applications upon request about available service capability features (discovery). For example, this mechanism is applied when installing or upgrading a service-capability server.

The following describes the basic mechanisms between application server and service capability server.

- Request of event notifications

  This mechanism is applied when a user has subscribed to an application and that application needs to be invoked upon receipt of events from the network related to the user. For example, when a user subscribes to an incoming call-screening application, the application needs to be invoked when the user receives a call. It will therefore request to be notified when a call set-up is performed, with the user number as called party number.

Once OSA basic mechanisms have ensured that an application has been authenticated and authorised to use network service-capability features, it is also important to handle end-user related security aspects. These aspects consist of the following:

- end-user authorisation to applications, limiting the access of end users to the applications to which they are subscribed;
- application authorisation to end users, limiting the usage by applications of network capabilities to authorised (i.e. subscribed) end users;
- end user's privacy, allowing the user to set privacy options.

8.8 Description of the API

As mentioned earlier, the OSA API is basically a cut-down version of the Parlay API with a few added extras. The API is made up of two basic parts:

- the framework part (this was explained in section 8.7);
- the service capabilities (being the services that operate across the API accessing the applications).

These service capabilities are as follows.

- Generic call control

  The generic call control SCF provides the basic call-control capabilities for the API. It allows calls to be instantiated from the network and routed through the network. The call model is based around a central call model that has zero to two call legs that are active (i.e. being routed or connected), each of which represents the logical relationship between the call and an address. However, the application does not have direct access to the call legs. Generic call control supports functionality to allow call routing and call management for CAMEL Phase 3 and earlier services.

  Generic call control is represented by the IpCallManager and IpCall interfaces that interface to services provided by the network. Some methods are asynchronous, in that they do not lock a thread into waiting while a transaction performs.

  In this way, the client machine can handle many more calls than one that uses synchronous message calls. To handle responses and reports, the developer must implement IpAppCallManager and IpAppCall.

- Generic user interaction and call user interaction

  The generic user interaction interface and call user interaction SCFs are used by applications to interact with end users. The GUI is represented by the IpUIManager, IpUI and IpUICall interfaces that interface to service capabilities provided by the network.

  The IpUI interface provides functions to send information to, or gather information from, the user, i.e. this interface allows applications to send SMS and USSD messages. An application can use this interface independently of other SCFs. The IpUICall interface provides functions to send information to, or gather information from, the user (or call party) attached to a call.

  To handle responses and reports, the developer must implement IpAppUIManager, IpAppUI and IpAppUICall interfaces to provide the callback mechanism.

- Data session control

  The data session control provides a means to control, on a per-data-session basis, the establishment of a new data session. This means, especially in the GPRS context, that the establishment of a PDP session (not the attach/detach mode) is modelled. Change of terminal location is assumed to be managed by the underlying network and is therefore not part of the model. The underlying assumption is that a terminal initiates a data session and the application can reject the request for data session establishment, can continue the establishment, or can continue and change the destination as requested by the terminal.

  The modelling is similar to the generic call control, assuming a simpler underlying state model. An IpDataSessionManager and IpDataSession object are the interfaces used by the application, whereas the IpAppDataSessionManager and the IpAppData Session interfaces are implemented by the application.

- Network user location

  The network user location (UL) SCF provides the IpUserLocationCamel interface, which provides methods for periodic and triggered location reporting. Most methods are asynchronous, in that they do not lock a thread into waiting while a transaction performs. In this way, the client machine can handle many more calls than one that uses synchronous message calls. To handle responses and reports, the developer must implement IpAppUserLocationCamel interface to provide the call-back mechanism.

- User status

  The user status (US) SCF basically provides the application with a means of finding out what the terminal is up to at any particular time. This means that it is able to find out when a user completes a call and when the user switches the telephone on or off.

- Terminal capabilities

  The terminal capabilities SCF enables the application to retrieve the terminal capabilities of the specified terminal. For Release 99 this can only be done for WAP users. The applications to make use of these capabilities are many and varied; however, the following section shows a couple of examples to explain the operation.

8.9 Using the API

To describe more clearly its use, a couple of examples of how the API works in practice are included here (see Figs 8.6 and 8.7). The sequence diagrams typically show the objects that reside on either side of the interface and the information flows (methods) that flow across the interface. The dark boxes represent the objects on the application side and the lighter boxes those that reside on the network side of the interface.

8.9.1 Generic Call Control Service — Alarm Call

The sequence diagram in Fig 8.6 shows a 'reminder message', in the form of an alarm, being delivered to a customer as a result of a trigger from an application. Typically, the application would be set to trigger at a certain time; however, the application could also trigger on events. The following sequence corresponds to the numbers shown in Fig 8.6.

1 This message is used to create an object implementing the IpAppCall interface.

2 This message requests the object implementing the IpCallControlManager interface to create an object implementing the IpCall interface.

3 Assuming that the criteria for creating an object implementing the IpCall interface (e.g. load control values not exceeded) is met, it is created.

4 This message instructs the object implementing the IpCall interface to route the call to the customer destined to receive the 'reminder message'.

5 This message passes the result of the call being answered to its call-back object.

6 This message is used to forward the previous message to the IpAppLogic.

7 The application requests a new IpUICall object that is associated with the call object.

8 Assuming all criteria are met, a new IpUICall object is created by the service.

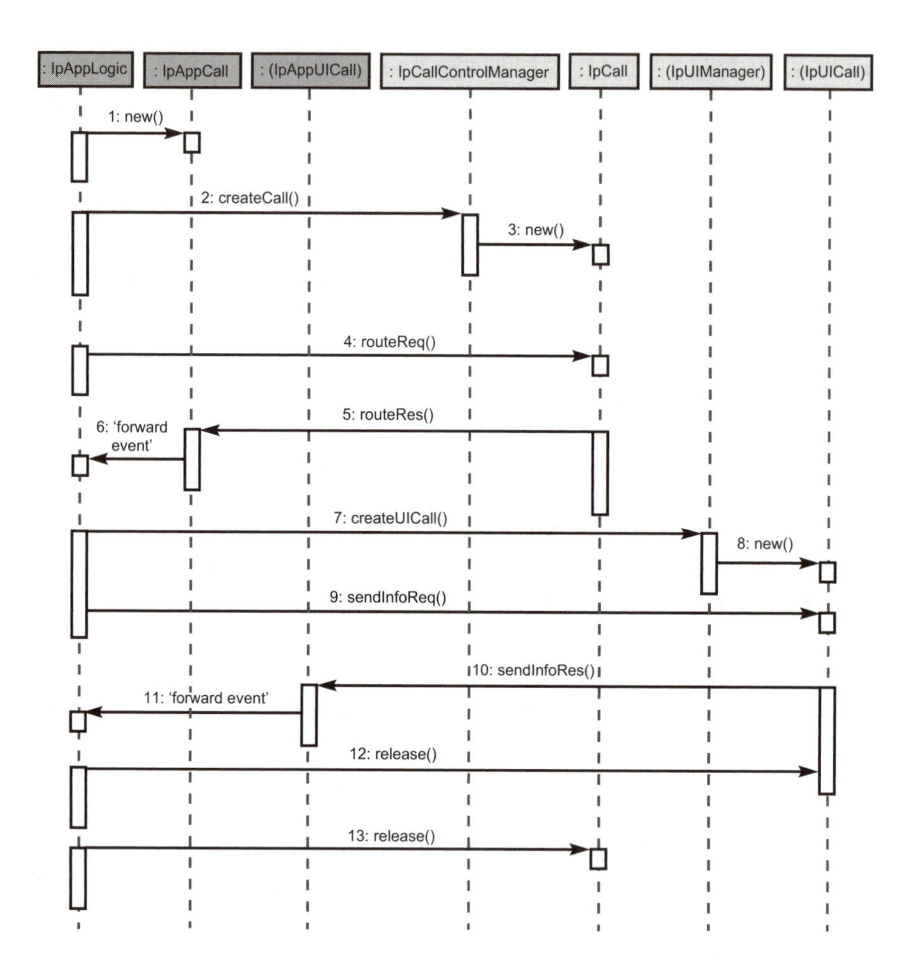

Fig 8.6 Sequence diagram for an alarm call.

9 This message instructs the object implementing the IpUICall interface to send the alarm to the customer's call.

10 When the announcement ends, this is reported to the call-back interface.

11 The event is forwarded to the application logic.

12 The application releases the IpUICall object, since no further announcements are required. Alternatively, the application could have indicated P_FINAL_REQUEST in the sendInfoReq, in which case the IpUICall object would have been implicitly released after the announcement was played. The application releases the call and all associated parties.

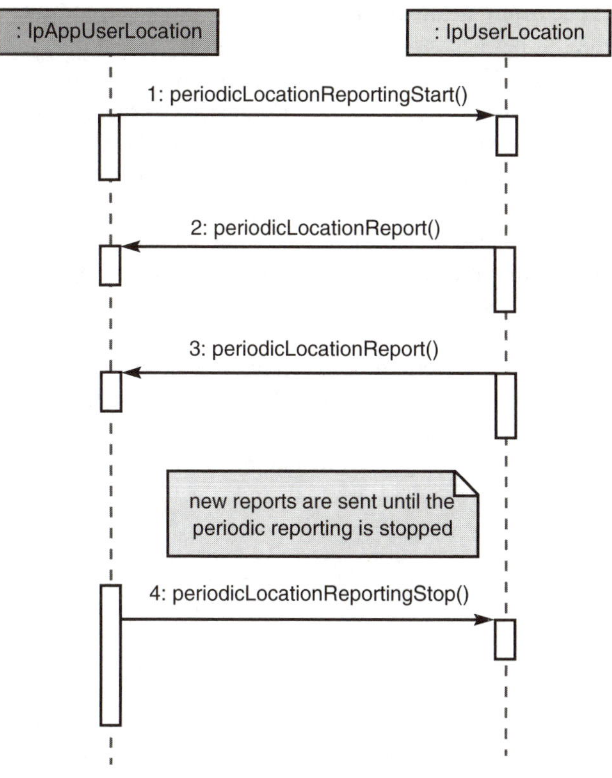

Fig 8.7 Sequence diagram for a periodic request.

8.9.2 User Location Interrogation — Periodic Request

The sequence diagram (Fig 8.7) shows how an application requests periodic location reports from the user location service (the following sequence refers to the numbers on the diagram).

1 This message is used to start periodic location reporting for one or several users.

2,3 This message passes the location of one or several users to its callback object. This is repeated at regular intervals until the application stops periodic location reporting (see next message).

4 This message is used to stop periodic location reporting.

8.10 The Future of OSA

So what does OSA offer for the future? OSA has been developed in conjunction with ETSI SPAN 3 and the ITU-T with the express purpose of ensuring one consistent API across the board. In other words there is only one international recommendation in the ITU-T and one standard in ETSI and they are exact mirror images of each other. This means that the one API for access to third party service provision can be implemented in many different ways whether it is to satisfy the target of supporting a virtual home environment as in the OSA case or to satisfy fixed network requirements which was the major goal of ETSI. This fundamental requirement is set to proceed for future developments.

OSA is in the process of defining the requirements for 3GPP's Release 4/5 specifications and, in doing so, has set up a joint working group with ETSI and the ITU-T (as before). However, this time they are including Parlay and JAIN in the talks. JAIN has also been developing a similar API for the Java community. As JAIN wanted to ensure no overlaps or differences in the call control part of the API, a number of changes to both the OSA Release 1999 specification and the JAIN specification held up proceedings until these could be sorted out. The group wants to ensure that this does not happen again and therefore will work on at least the call control part together. Release 4/5 is set to include the following additions.

- User location:

  — integrating location services with geographical positioning.

- Terminal capabilities:

  — Release 99 was only applicable to MExE and WAP telephones, and therefore needs enhancing to include GSM;

  — security mechanisms for display of terminal capabilities information.

- Enhanced user profile management:

  — integrating PSEM with network and framework SCFs.

- Enhanced session control:

  — providing enhancements to report QoS when negotiated or changed.

- Charging:

  — supervise user activities for on-line charging features;

  — allow applications to access the account;

  — allow applications to add charging information to network-based accounts;

  — inform applications on network-based charging event.

- Security:

  — guarantee secure access to user-confidential information.

8.11 Conclusions

This chapter has set out to introduce to the reader the concepts of OSA. In the process the architecture concerned has been illustrated, giving examples where appropriate. It has been shown how the OSA group, right from the start, has ensured collaboration with international standards forums to ensure one international specification for an API accessing third-party service offerings, possibly outside a normal operator's domain. It has also been shown how the group intends to continue this collaborative process in the study of the 3GPP's Release 4/5 specifications, and what the future offers the telecommunications industry.

The developments under construction in the joint OSA group will play a major role in the integration of the fixed network IP offerings and those of mobility, providing that seamless environment we all so desperately need.

9

SERVICES VIA MOBILITY PORTALS

D Ralph and C G Shephard

9.1 Introduction

Mobility portals look set to become the window through which the user will access a whole range of innovative services. Wherever and whenever, the mobility portal will warn you, inform you or just entertain you.

There is a distinction between the mobility portal and the mobile portal. The concept of mobility extends to include terminal independence through user profiles, additional value-add services such as 'find me, follow me' call routing, and integration with existing systems in the fixed network. The mobile portal, however, is being expressed as a cut-down version of existing Web-based applications, such as news, weather and e-mail.

The explosion in use of the short message service (SMS) for sending text messages demonstrates a natural evolution path to wireless application protocol (WAP) services on smartphone and personal digital assistant (PDA) devices. Support from many vendors has created an environment where the applications delivered through the mobile portal will increasingly substitute access through the fixed network.

However, although the Internet and mobile telephones have independently been successful (see Figs 9.1 and 9.2), it does not always mean that combining these technologies will also be successful. Evidence of this is clear in the combination of TV and telephone technologies, both successful in their own right, but the uptake of personal videophones has not happened. Mobile videophone is considered a killer application. However, as will become clear from this chapter, to achieve a high take-up is unlikely in the foreseeable future due to bandwidth limitations and device battery life. A more likely scenario is that a short video clip downloaded in non-real time could be played back to the user to provide visual information.

In addition, users may subscribe to specialised information feeds, for example, for stock trading, weather, ski and snow conditions, and horoscopes.

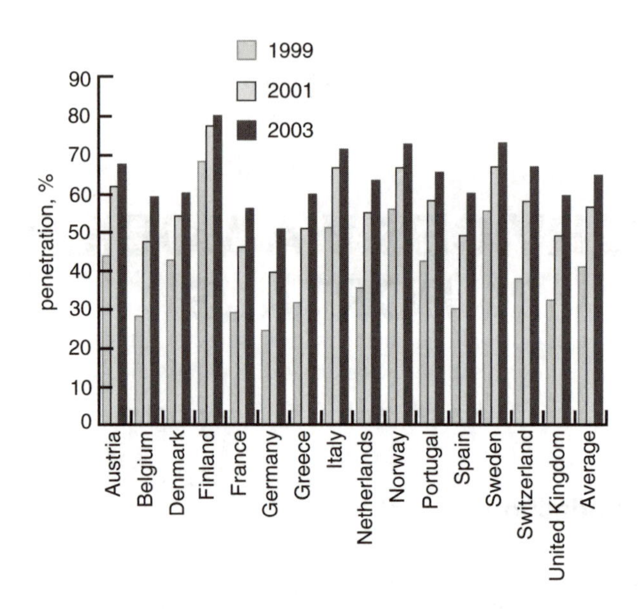

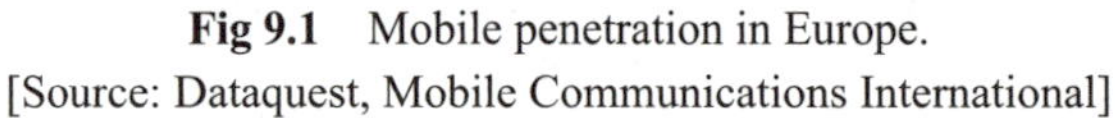

Fig 9.1 Mobile penetration in Europe.
[Source: Dataquest, Mobile Communications International]

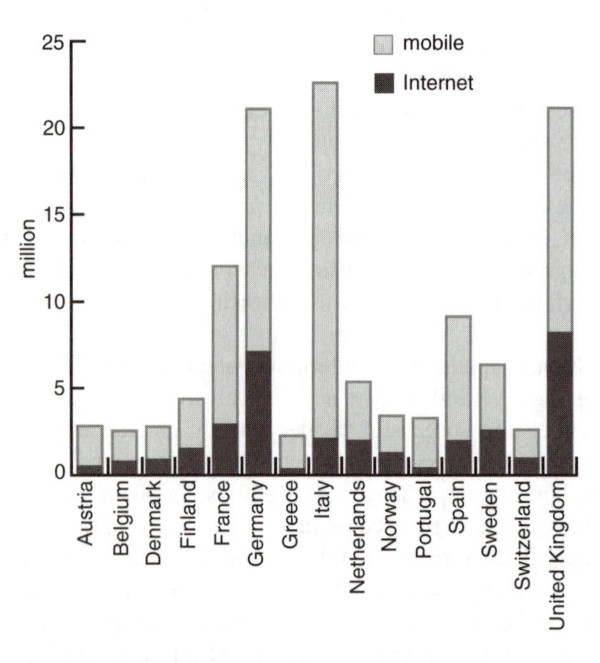

Fig 9.2 Internet and mobile subscribers in Europe 1998.
[Source: Dataquest, Mobile Communications International]

Finally, the services will offer specialised alerting through partners such as auction, travel sites, banking, and messaging providers.

9.2 The Rise of the Mobility Portal

Without doubt the explosive growth rate of Internet users and the significance of accessing content through the use of Web-based portal services will continue. While this highlights the demand for content from the fixed network, it cannot be assumed that the same type or level of demand will be present in the mobile environment. The differences in network and device capability will require a different solution in providing content through the mobile portal.

In assessing why the mobility portal is important, it can be seen from looking across the value chain that a number of key elements are available to be exploited in delivering new mobility services:

- the existing infrastructure will support these services, both in terms of core IP networks and content availability — whether this is from the customer's Internet service provider (ISP) or other content providers, it will enable the rise of the mobility portal;
- the prospect of 'always on' packet-based services delivered over 2.5G and 3G networks, such as GPRS and UMTS, will enhance the user experience in the way content is delivered over the next generation infrastructure provided by the mobile operator;
- the first generation consumer equipment is readily available and supports basic Web access and WAP services — this is in the form of smartphone and PDA devices;
- in the short term, existing content can be redeveloped for delivery to the smartphone or PDA device, which will enable the rapid deployment of mobile services — although this will only be through provision of existing generic Web portal services on the mobile device, it will include search engines, personalisation, snap-shot text information, and basic messaging.

It is the very issue of content and what services the user will require that will determine the successful uptake of service access through the mobility portal. Figure 9.3 illustrates some of the anticipated benefits from a wireless portal.

Questions have to be answered to determine the customers' needs when mobile — what do they want, when, why? Providers of mobility portals will need to assess customer requirements and actual usage to develop context awareness in order to present the most relevant content. If the network knows the weather conditions, user location, time and who the user is accompanied by, then this gives the network an opportunity to provide information ahead of the user's requirement for this information, further developing the mobile device as a life-style tool.

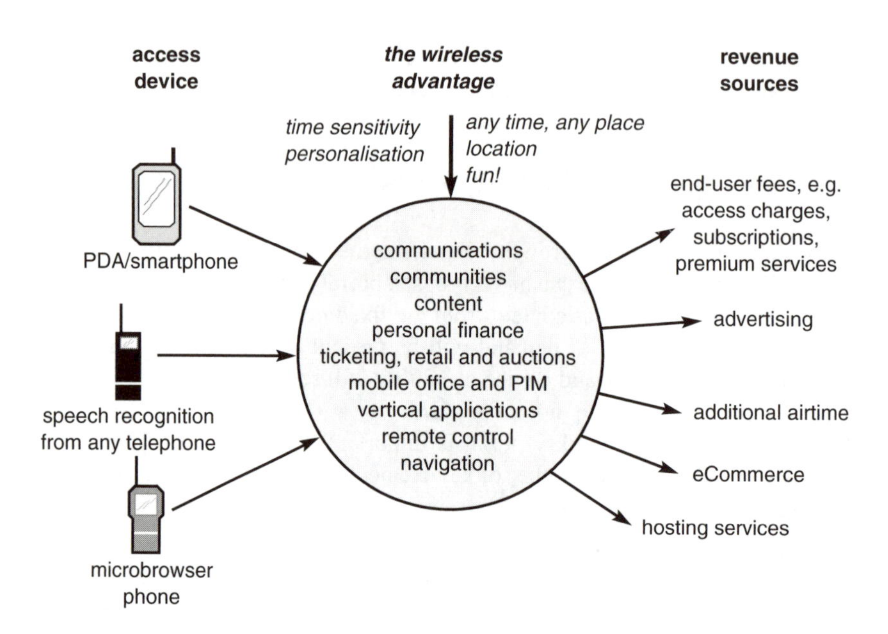

Fig 9.3 Benefits of a wireless portal.

The user's expectations must be managed at this early stage in the adoption of these new technologies — there is a significant amount of hype surrounding the quality and quantity of content that is available.

Future applications need to develop the unique attributes of the mobile device. The opportunity for the network to provide location-based services, and the personal nature of the mobile device lend it to acting as a payment device.

9.3 Inside the Machine

The current UK digital mobile network has evolved from the first generation analogue network and is based on the global system for mobile communications (GSM).

The current key technologies supporting mobility include:

- cellular;
- paging;
- cordless (radio and infra-red);
- mobile satellite;
- in-flight systems;
- wireless LANs (IrDa and radio);

- mobile terminals;
- fixed network mobility;
- professional mobile radio (PMR)/mobile data;
- mobile middleware;
- emerging third generation mobile systems (3G mobile).

Network technologies are evolving towards the 3G mobile vision; this will require an evolution through extensions to the existing 2G mobile network technologies. In the near future this will involve:

- GSM component technologies;
- SMS and cell broadcast;
- GPRS (general packet radio service);
- EDGE (enhanced data rates for GSM evolution);
- UMTS (universal mobile telecommunications system);
- Bluetooth;
- GPS and other location-based techniques (mobile positioning protocol);
- mobile agents and intelligent networks.

The next generation mobile system will also rely heavily on emerging application technologies such as:

- wireless application protocol (WAP);
- XML/XSL (extensible markup language);
- Java technology (Java 2 micro edition);
- SIM application tool-kit;
- lightweight efficient application protocol (LEAP) [1];
- compact HTML.

It is clear that the Internet will play a pivotal role, requiring increasing quality of service (QoS) and allowing better integration of applications across different terminals and bearers, probably using APIs to provide access to functionality.

The importance of voice must not be underestimated as a mechanism for controlling services by natural language commands, for example, to add an appointment to your calendar, or, more simply, as an alternative to typing a response to an e-mail. Voice browsing of a mobility portal will be standardised through the use of VoiceXML and provide a development environment to deliver powerful niche applications [2].

Before service providers can offer content or applications using mobility portals, a number of issues require further consideration.

9.3.1 Wireless Application Protocol

WAP [3] is a specification for a set of communication protocols to standardise the way that wireless devices, such as cellular telephones and radio transceivers, can be used for Internet access, including e-mail, the World Wide Web, newsgroups, and Internet relay chat (IRC).

While Internet access has been possible in the past, different manufacturers have used different technologies. In the future, devices and services that use WAP will be able to interoperate (see Fig 9.4).

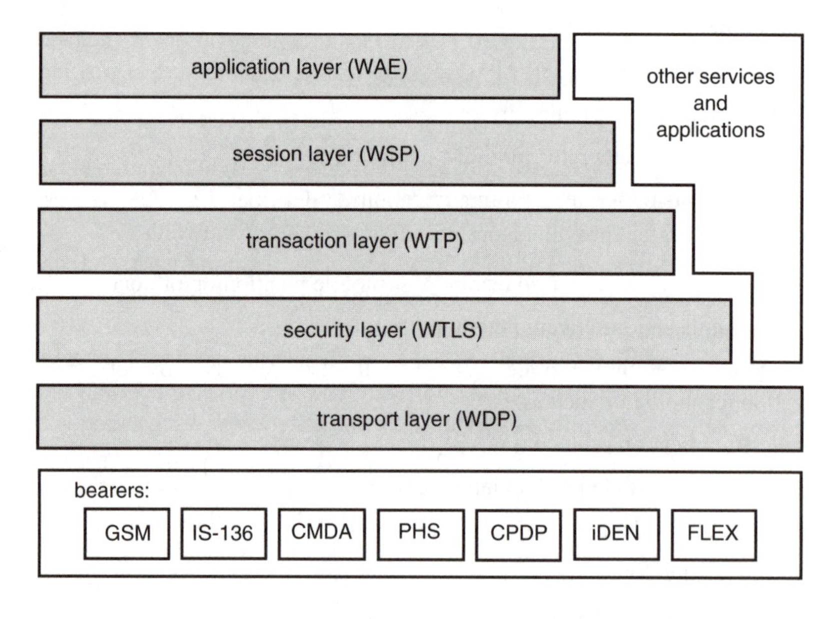

Fig 9.4 The WAP protocol stack.

The WAP layers are:

- wireless application environment (WAE);
- wireless session layer (WSP);
- wireless transport layer (WTP);
- wireless transport layer security (WTLS);
- wireless datagram protocol (WDP).

Following independent development, four companies proposed the WAP standard — Ericsson, Motorola, Nokia, and Unwired Planet (which is now Phone.com).

The popularity of WAP in Europe [4] must be contrasted with the developments in Japan where the i-mode service has significantly better functionality. The service uses compact HTML (cHTML) and is provided over the Personal HandyPhone System (PHS) which is a packet-based 'always on' service. The use of cHTML permits colour graphics to be displayed although it still has limitations similar to the wireless markup language as used in WAP.

There are many reasons for the success of i-mode. It relies on open technology, allowing any Internet site to join in, content is written in a simplified version of HTML (cHTML), and no fees are charged for placement on the i-mode portal. The services are positioned as a unique mobile service and not as 'the Internet on your phone', thereby avoiding unfavourable comparisons with service from the fixed Internet. Also the low penetration of PCs in Japanese homes probably helps control user expectations. Finally, NTT DoCoMo has concentrated on growing core revenues from airtime usage and has chosen not to support the service through advertising or transaction revenue.

Mobile Internet services have now been introduced by rival operators to compete with i-mode. However, they have so far failed to attract significant numbers of subscribers or content providers, despite offering greater bandwidth.

By being first to market i-mode may now have an important head start and competitors offering WAP-based services cannot tap into the wealth of available i-mode content because of format incompatibilities.

'i-mode' is a brand name for NTT DoCoMo's Internet service. The technology and service offerings behind i-mode will continue to evolve to take advantage of improving network capabilities, including the introduction of 3rd generation mobile networks. In addition, DoCoMo has plans to roll out the i-mode service in other countries, particularly in mobile networks where NTT owns a share; one example is the co-operation with Hutchinson in Hong Kong. At the same time, DoCoMo is active in the WAP Forum, World Wide Web Consortium (W3C), and Internet Engineering Task Force (IETF), whose standards, in combination, will map the way forward for a globally compatible mobile Internet.

The use of transcoding engines (described as the HTML filter in Fig 9.5) is being proposed as a 'quick win' in providing services from existing Web content to the first generation of wireless terminals. This has met with limited success due to the complexity present in much of the multimedia-rich Web content. The use of Javascript, frames and image maps causes, in many cases, a failure to transcode the HTML into a suitable WML page that can be displayed on a smartphone.

The deployment of global mobility portals are essential to support the global traveller — they need to support the roaming capabilities of GPRS and GSM, and remove the added complexity for the user to change profiles to connect to the same services in different countries. The geographic distribution of network elements and the seamless access to services across different countries will strengthen the mobile operator's position as a service provider.

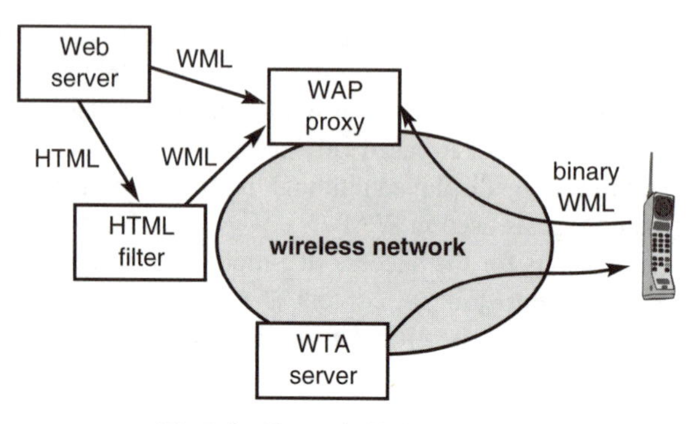

Fig 9.5 Example WAP network.

9.3.2 Terminals

While the 'look and feel' of the mobile device will have an impact on the user's perception of services, the usability of the device will have an impact on the serviceability of any application delivered over that device. Regardless of the content the portal provides, the terminal usability will prove significant in the promotion of services through mobility portals.

The types of mobile device range from the smartphone complete with NaviRoller, to a pen-based PDA device (see Fig 9.6), up to a subnotebook with keyboard and 1024 × 480 pixel display. Each of these devices has different attributes related to:

- screen size (resolution);
- input method — keyboard, touch screen, handwriting recognition, etc, through to the presence or otherwise of a speaker and microphone.

Fig 9.6 Next-generation PDA device with pen and speech based input, integral GPS and GPRS connectivity.

The multiple methods for input and display are compounded by the variable and often low bit rate of mobile networks. In order to provide suitable content, device characteristics will need to be available to the service provider to allow appropriate formatting and display to the range of devices outlined. The W3C have a working group on device capability and profile called Composite Capabilities/Preference Profiles (CC/PP); it is driving standards efforts to implement device profiles that will enable content providers to determine device capability.

Further proposals from the Salutation Consortium, Jini, and Universal PnP (UpnP) [5], are intended to deliver service discovery and device recognition capabilities [6] to enable the network to seek the services that the user requires and that are suitable for the terminal capability. For instance, the user may require a print service, but only have available an IrDa port through which to communicate with the printer.

The limitations of device capabilities in terms of storage and access to information will increase the need for additional support from the network. This may be provided through network-based back-up and synchronisation services or server processing for voice recognition to authenticate access to a service.

9.3.3 Addressing

It has been a concern for some time that the lack of IPv4 address space would prevent connections to the Internet. Several interim solutions have been implemented successfully. These include IP virtual private networks (IP VPNs) and classless Internet domain routing (CIDR). This problem will only become worse with an explosion of mobile devices that have an 'always-on' connection to the Internet. Ultimately the proposed migration to IPv6 with its 128-bit address scheme will remove this problem with available address space for devices connected to fixed or mobile networks. Forecasts report the size of the IPv6 address range will allow 10^{20} addresses (or one IPv6 address for every square inch of the earth).

Each mobile device that a user carries, or interacts with, will be uniquely identified by an IPv6 address and may well communicate to the user using Bluetooth devices connected within a 'personal bubble' or to an external network using UMTS. It is estimated that 10 billion Bluetooth addresses alone will be in use [7].

The domain name service (DNS) which is used to map a URL or e-mail address to a particular host/server may suffer. The explosion of devices will place significant load on the DNS infrastructure. However, given its distributed and scalable nature, this should not cause any significant difficulties.

The explosion of addressing may cause network integrity issues if the address range for IP or Bluetooth is incorrectly assigned. The globally unique media access control (MAC) address is managed by the IEEE and there are many examples of problems with unauthorised address use.

The home environment is where a significant portion of this addressing will be required — in particular, the ability for devices around the home to communicate, to share information on whereabouts, and to anticipate the family's needs. This could be extended to the fridge ordering fresh milk from the supermarket upon your arrival home from holiday. The use of cell broadcast or Bluetooth may be used to transmit special offers to a mobile user while walking past a specific shop with which they have registered an interest.

9.3.4 Billing

Getting information to the handset of a user is the simple part. Devising a mechanism that meets the commercial aspirations of all the players involved in making this happen is much harder.

Many network operators will partner with a publisher to manage the day-to-day operations of their portal. Service providers will want recompense for the information provided.

Even in the simple case of portal provision, a number of revenue models have been developed to describe the different ways in which the value can be expressed. For example, a simple revenue share might be appropriate for a traffic-congestion report, the operator charging the user 'per click' for information and passing on a share to the information provider.

A service offering sports news (for instance, all the latest scores) might require a subscription from the user to the operator, but a volume-dependent fee to the service provider. Other services would require more complex interactions between companies, for example, in the case of a car-hire scenario, requiring co-operation between an operator, an airport, an airline and a car-hire company. Resolving the issues of ownership of the service, software, and access to customer information is more challenging than building the basic service.

Initially, WAP services are expected to be expensive to use since the tendency is to be on-line for a long circuit-switched-data (CSD) call, since features such as interactivity and selection of more information are used by the end user. It takes several short messages to send one piece of information through WAP. Without specific tariff initiatives, there are likely to be some 'bill-shocked' and surprised WAP users when they see their mobile telephone bill for the first time after starting to use WAP, even if their user experience was good.

9.3.5 Security

There are several areas of WAP security that will need additional development to ensure the end-to-end security requirements for transactions involving payments or instructions requiring non-repudiation. These include what is being called premature encryption end-point. The WAP gateway acts as a proxy and decrypts the incoming WTLS session and encrypts as SSL3.0 for the outgoing connection to the origin server. There is an opportunity on the gateway to store as plain text the information passed between the two end-points. Developments are under way to allow session redirection to a gateway within the firewall of the bank or service provider so that they can ensure control and security of the session.

The WAP specification (v1.1) supports the use of certificates in a way similar to X.509 and encryption is supported at 56 bit and 128 bit. In the future, the WAP specification will be extended to include the wireless identity module (WIM) which may be a replacement for the SIM card, containing a personal digital signature to identify and authenticate the user. The addition of a cryptography library will enable the developer of WAP services to provide a range of functions that allow the user to digitally sign and encrypt a message before it is sent from the mobile device.

9.3.6 Interoperability

The provision of mobile data services in the past has always been via proprietary technologies [8]. The standardisation of WAP allows developers and services to be less restricted in their range of mobile networks or mobile devices. The first release of gateway and device products has the inevitable problems associated with new technology and significantly more testing will be required between different vendors' products. This is an area where the service provider can add value in providing an integrated service to an agreed level of quality and compatibility with other services.

In future, care must be taken that proprietary extensions do not confuse and frustrate users because their device does not support an extension specific to a single mobile device. On the issue of browser upgrades, the telephone manufacturers are not prepared to commit to a policy that would allow the user to update the telephone software — instead they may be forced to purchase a new telephone [9].

9.3.7 Quality of Service

The issue of quality of service (QoS) has plagued IP networks since they became commercialised. Several IETF standards have been proposed to satisfy the user requirements — these include resource reservation protocol (RSVP) with IETF integrated services (IntServ), differentiated services (diffserv) and multi-protocol label switching (MPLS) [10]. The RSVP protocol aims to deliver a service level agreement to the user at the access to the network. The differentiated services protocol is employed between core networks in order to meet the minimum service level agreements for the end users. MPLS, a flow labelling protocol, has recently been introduced in an attempt to obtain end-to-end QoS for IP sessions. This has some of the attributes of an ATM cell and allows very fast switching of packets, labelled for a particular flow type through a particular route.

Other standards have been adopted by the network equipment manufacturers, examples being weighted random early drop (WRED), and network layer routing.

The WRED system is crude, in that it drops packets not considered to be priority once its internal queues become full. Network layer routing examines details such as port number and protocol type. For example, RTP/UDP carrying traffic on a port used for voice over IP will be given priority routing at each hop through the Internet.

These issues will primarily affect the core network and to a certain extent be solved by the above protocols. However, there is still no support for an end-to-end guaranteed bandwidth IP session.

The mobile network suffers from QoS issues, such as dropped connections, lack of resources, and reduced bandwidth, due to too many users sharing the available capacity. These factors will significantly restrict the availability of mobile services that compete directly with fixed network services, such as videoconferencing or high-speed file transfer.

While it is true that 2 Mbit/s may be available using UMTS by the year 2010, this will only be available for a single user to watch a high-resolution video. The issue of access network congestion may well have an impact on network design if urban 'hot spots' demand increased bandwidth.

Some of the issues over dropped connections are being resolved by the use of soft handover in the next-generation networks. This occurs when two base-stations are tracking the mobile user and, as the signal strength moves from one station to the other, so does the handover. The network may be able to retransmit packets lost due to the mobile user's signal strength being temporarily lost. However, if the user enters the channel tunnel on Eurostar, the only solution to maintaining a connection will be to fit pico-cells to the carriage.

As users are demanding higher levels of QoS from IP networks for fixed services delivered on these networks, so too will the mobile user. The demand for real-time services may be sufficiently low to cause less difficulty — video-streaming can use buffering to delay the transmission. Videoconferencing cannot accept high network latency, but it may be that this is not the 'killer application'.

9.4 Future Developments

The roadmap for WAP includes integration with existing mobile technologies from the mobile execution environment (MExE), SIM application tool-kit, VoiceXML and location positioning. In the immediate future, the WAP specification (from 1.2 to 1.3) will evolve to include 'push' services and a wireless identity module.

The advent of 'always-on' GPRS network connectivity will support the proposed 'push' method — this will enable the update of information to a user independent of their need to request it.

The WIM will support the requirements to provide authentication of a user when committing to a transaction, particularly important for banking and mobile commerce applications.

While aimed at wireless operators and their wireless application protocol telephone customers, the mobility portal will know the location of users and provide location-specific content. Users will benefit from the creation of preference profiles and data will be deliverable either upon request or as determined by their schedule. The user profile can reflect common locations, such as home, at work or at a branch office.

The other protocols, such as SIM application tool-kit and MExE, are widely supported and may be consolidated with the WAP standard to provide a common development environment. The use of XML as a content storage format will potentially meet the needs of adapting and delivering appropriate information to the user.

The commercial pressure to be first to market will provide challenges in managing user expectations of WAP services. It may have been better to wait until WAP 1.2 and GPRS are implemented before launching WAP services commercially. Plans for updating the WAP 1.1 telephones, that are now being sold, to WAP 1.2 are unclear.

It is certain that WAP will be important in enabling the smooth transition from one bearer to another, such as the migration of existing applications to GPRS. Both developers and users are at a crossroads in mobile communications as we move from voice to non-voice centric services. Both development and marketing teams have much to learn from WAP and other services. Within 18 months, all new mobile telephones will support WAP, and, given the huge subscriber growth in Japan, there will be many subscribers.

Presented below is a list of areas that will need to be overcome to enable WAP services to reach critical mass:

- WAP 'push' support on packet-based mobile networks (differentiated from existing SMS services);
- wireless telephony architecture (WTA), not implemented;
- wireless identify module, yet to be integrated;
- premature encryption end-point;
- small downloadable unit size (1400 bytes maximum);
- WDP reliable connectionless datagram protocol, not implemented.

Alternatives such as CHTML and other existing and developing Internet standards may provide the solution to these omissions in WAP, given the predicated bandwidth on next generation mobile networks and increasing device capability.

The device capabilities will increase significantly with developments to produce a lower power, faster speed CPU such as the crusoe chip. Future improvements in battery life and the addition of peripherals through the use of a sleeve that slides on to the PDA will expand the possibilities for delivering services to the mobile device. The use of client-side processing to enable preprocessing for voice recognition software will allow a reduction in bandwidth requirements across the mobile network. Improvements in algorithms for voice control may see this become an important input mechanism.

9.5 Conclusions

The future developments of the mobility portal must embrace the entire range of existing Web-based content if it is to succeed as the default method of access to the network. This will include all applications from gaming to mobile commerce.

Services should be designed so that the relevant elements of a service are available over the appropriate client device, the requirement being that the network recognises the device capability and network connection characteristics.

There will not be a single 'killer application' for the mobile device although its unique attributes including location positioning and personal nature provide an opportunity for the mobile device to become the portal through which the user interacts while 'on the move'.

It is highly likely that WAP will be superseded within the next 3-5 years by the evolution of existing Internet standards; the requirement for legacy support will remain for WAP-enabled mobile telephone devices. The reasons for the limited lifetime of WAP are:

- low bandwidth;
- high latency;
- low-resolution monochrome displays;
- dropped calls and other quality-of-service issues;
- low processing power.

These are not limitations to the present WAP service, but once these areas have been improved or overcome, the tendency will be to use existing Internet standards end-to-end, as opposed to the WAP architecture of creating a wireless version of these standards. Even if successful, once all devices and networks support WAP, it will cease to differentiate and therefore will not reduce subscriber churn. However, it will provide the user with a convenient access to services whenever and wherever required.

WAP is an interim solution to the scarce availability of radio bandwidth and the lack of quality screen 'real-estate' available on current devices. The limitations such as bandwidth, screen size, battery life and processor speed will all improve, enabling devices in the future to support TCP connectivity and allowing them direct access to HTML. Many devices, including the smartphone, are already supporting HTML and WML implemented in the browser technology.

With so many heavyweight carriers, equipment manufacturers, and software and applications developers backing the Wireless Application Forum, a momentum is being generated that will drive WAP-based equipment and services forward into the future. The first WAP devices are already being distributed and a number of WAP compatible services have been announced. WAP has arrived and is tangible. People can now access the Internet and other data services via mobile telephones and other handheld devices.

With operators around the world undergoing a paradigm shift, from telephone company to value-added information broker, there is a clear understanding that future success in communications markets will require a holistic approach to the rapidly converging worlds of telecommunications, computing and multimedia. This means developing advanced application and content packages to be sold alongside basic telecommunications services. WAP acts as a facilitator for such an approach to mobile communications.

Despite the drawbacks discussed in this chapter, WAP certainly seems to be shaping up to play a major role in facilitating the 'brave new world' of the personalised mobility portal. WAP is a powerful tool — it enables 'any time, anywhere' connectivity to a wealth of information, whether for leisure or business applications.

This chapter has demonstrated, through the discussion of different mobility portals, some of the problems involved in providing content, applications and filtering information to a new generation of wireless devices.

References

1 The lightweight and efficient application protocol (LEAP) Manifesto — http://www.freeprotocols.org/leap

2 '*Voice portals: Ready for Prime Time?*' (July 2000) — http://www.zdnet.com/anchordesk/stories/story/0,10738,2601898,00.html

3 WAP Forum: '*WAP architecture specification*', (April 1998) — http://www.wapforum.org

4 '*European wireless portal use to boom*', (July 2000) — http://www.allnetdevices.com/industry/market/2000/07/07/ european_wireless.html

5 UPnP, Jini, Salutation — http://www.cswl.com/whiteppr/tech/upnp.html

6 Service discovery and management — http://www.salutation.org

7 '*Bluetooth SIG adds protocol*', (July 2000) — http://www.allnetdevices.com/developer/news/2000/07/10/ bluetooth_sig.html

8 Johnston, B., Fenton, C. and Gilliland, D.: '*Mobile data services*', BT Technol J, **14**(3), pp 92-108 (July 1996).

9 Future phones — http://www.FutureFoneZone.com/whitepaper.htm

10 Spraggs, S.: '*Traffic engineering*', BT Technol J, **18**(3), pp137-150 (July 2000).

10

TERMINAL DEVELOPMENTS AND THEIR MEDIA CAPABILITIES

W Johnston

10.1 Introduction

There are now more mobile telephones in the UK than fixed lines and the number of PDAs is increasing rapidly. Customers buy a telephone for the looks or the features and often do not care who the operator is. Mobile devices are the gateway to all telecommunications networks and services and can consequently be used to increase customers and therefore generate revenue for a network operator. With so many services and applications being promised for 3rd generation (3G) networks, operators now face the challenge both of installing the network capability and of procuring terminal devices that aim to match at least some of these expectations.

Figure 10.1 shows an end-to-end representation of a network. Traditional network operators spend a lot of time and effort on developing and implementing networks and ensuring they are compatible both internally and with other licensed operators (OLOs). Generally, there has been less in-depth involvement with service development and integration and very little involvement with terminal development/manufacture/implementation. Given the complexity and flexibility of 3G, this situation must change if 3G operators are to develop an optimised solution. BT recognises the importance of the terminals and, consequently, is scoping its requirements with all the world's device manufacturers to fulfil the 3G requirements.

The word 'terminal' can mean anything from the fixed telephone on the desk to set-top boxes or an integrated handheld cellphone with organiser. In order to bound this discussion, in this chapter terminals will relate only to wireless-enabled mobile devices that can be attached to a cellular network. This is still a very large area that is ever changing, e.g. in some parts of Europe mobile telephones are becoming fashion accessories and in Japan mobile devices are becoming so small that they are incorporated into jewellery/children's toys and watchphones with built-in cameras.

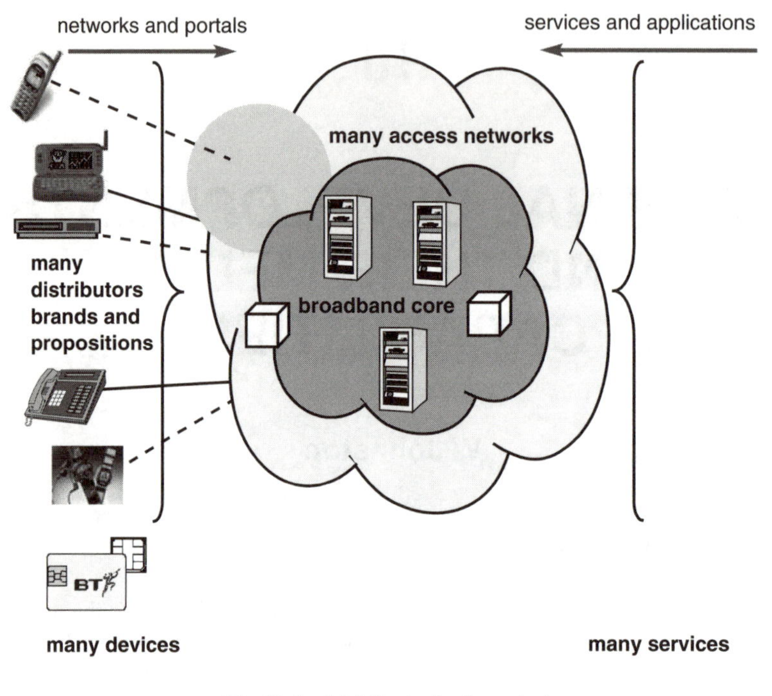

Fig 10.1 Mobile device boundaries.

Given these trends, this chapter will summarise some of the major components and technologies currently being considered and where these technologies might go in the future.

10.2 Fundamental Requirements

There are some things that are fundamental to a mobile device and radio capability is one of them. This may sound obvious but not all operators will win a UMTS licence and in this case there may be a greater requirement for enhanced services such as GPRS and/or EDGE on the mobile device. Consequently BT has to try to accommodate as many joint venture requirements as possible in terms of capabilities supported by any one handset, leading to lower costs from economies of scale.

10.2.1 Radio Frequency

During network roll-out of various network/transmission technologies such as EDGE/GPRS and UMTS across BT's joint ventures, it is expected that terminals will support some or all network technologies. This will maximise network coverage and operator revenue and provide a type of mobile seamless service handover between the different network technologies. Requirements for at least the first three years of UMTS lifespan are for a combined WB-CDMA/GSM 900/ EGSM/GSM1800 terminal. Additional requirements include GPRS, EDGE and TDD depending on the country and operator, since not all operators will win a UMTS licence. Even in countries where a UMTS licence is acquired, it will take a finite time to roll out a nationwide network, and multiband telephones will enable a smooth transition for customers and operators alike.

The networks support both circuit-switched and packet-based services and it is assumed that most devices will support both types of connection within one device. Some advanced devices will support simultaneous circuit and packet services enabling, for example, both voice and data services to be instigated together. This will have a major impact on the way people use their mobile devices in the future.

10.2.2 Device Format

There has been much speculation regarding the way the terminal looks. Bearing in mind that there will be a large global installed base of laptop computers, cellular telephones and interconnections between the two, it is safe to assume that mobile device formats will evolve gradually as technology progresses. There will not be a 'Star Trek' style communicator available from day one of launch. A number of device/product formats are expected to be available (see Fig 10.2) to enable different market opportunities to be addressed:

- voice-centric terminal (similar to current mobile telephones) — this terminal will primarily be used for voice, but be able to link to PCs or PDAs, with connectivity types likely to be Bluetooth/USB/FireWire/IrdA/RS232;
- integrated PDA and telephone (mainly PDA with telephone features);
- communicator type device (more integrated than above);
- data card for PC.

As new features such as video download become increasingly popular, mobile devices will change in style and function to support the new features. This then leads to the question of how the video clips will be viewed.

10.2.3 Displays

Displays are an important part of any device dictating the size of the device and consuming a relatively large quantity of total power. Given that one of the promises of 3G is the capability to download video clips, there is an added incentive to

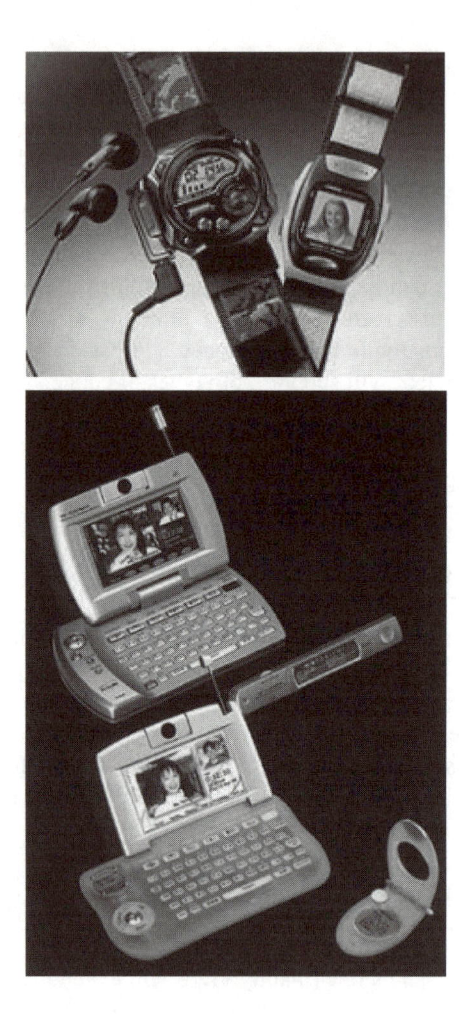

Fig 10.2 Examples of mobile devices.

implement colour screens that require even more power than current LCD. Since screens can be a limitation to many functions required for UMTS, even Web browsing, it is worth taking a quick look at some of the technologies available. Most handsets and PDAs use liquid crystal display (LCD) that can be classified as:

- AMLCD (active matrix LCD);
- PMLCD (passive matrix LCD).

AMLCDs provide the most comprehensive display capabilities whereas PMLCD is more basic but the power consumption and the manufacturing costs are lower.

Some of the PDAs on the market use more sophisticated displays, but this is at the cost of power consumption and hence battery life. In these cases the technology used is called TFT LCD (thin film transistor LCD) and is an active matrix LCD. These LCDs can display up to 65 536 colours, but, as they are active, their power consumption is much higher. They quite often require backlighting as well.

10.2.4 Where Next?

There are some new emerging technologies such as projection systems contained within a cellphone antenna that can display an image on to a nearby wall or flat surface. Another technology uses an LED/laser system that is the size of an LED and can write a VGA image on the eye. The advantage here is that the mobile device could be the size of a ball-point pen but still enable the user to see a high-resolution image. A more conventional technology is light-emitting polymer (LEP) which is a flat-panel colour display technology. The benefits are:

- high levels of brightness — they are self-luminous and do not require backlighting;
- low power consumption;
- thinner, more compact display than LCD;
- less weight;
- wide viewing angle (up to 160°).

LEPs can also be flexible offering the possibility of roll-up screens, again allowing the mobile device to be smaller with a pull-out screen to be used when required.

10.2.5 Power

Between the need for high quality, colour screens and applications that require more processing power such as multimedia, another major area to consider is power requirements and battery life. While BT is not actively involved in specifying this technology it is still important to understand how this is changing in the future. Battery technology has progressed quite slowly over the years. Nickel cadmium (NiCd) was the mainstay of power units for mobile devices. These were gradually replaced by nickel metal hydride (Ni-MH) offering greater energy densities and also being kinder to the environment. Energy densities were still quite low and the physical size of the power cells dictated the size and weight of the cellphone. Many cellphones today use lithium ion (Li-ion) which offers even greater energy densities and helps keep the weight down. Current lithium ion batteries contain a liquid electrolyte that must be contained within a solid reservoir. This limits the size and, more importantly, the shape of the battery in the same way as NiCd and Ni-MH. A further enhancement on this is Li polymer which uses a solid plastic electrolyte that can be shaped to (almost) any form. In addition, the lithium polymer batteries are lighter and safer than lithium ion designs. (Initial Li-ion batteries had a tendency to explode while charging.)

Another battery technology still under development is the fuel cell. Although fuel cells have been used for several years (in the space shuttle and currently being tested by the US Army for mobile radios), there are none available in a package small enough for a mobile device. A fuel cell operates by electrochemically combining fuel (e.g. hydrogen, methanol) with oxygen (O_2) from the air to produce electrical power. One benefit would be that the telephone could be 'charged/filled' within a few seconds and last many times longer than current battery technology. Interestingly, there is also a small amount of 'exhaust' composed of carbon dioxide and water vapour and companies working in the area reckon there is some 3-5 years of research to be done before anything is available in the mobile terminal space. Despite this, research models have produced cells with three times the energy density of Li-ion currently being used. There are other technologies, such as Zinc-Air, that provide high energy densities (in fact up to three times current technology), but they have characteristics that limit their use, i.e. they degrade in contact with air so they could be used in emergency cases but are not suitable for everyday use.

In summary, this section has briefly reviewed some of the technologies that have an impact on the physical characteristics of a mobile device. There are several other technologies, such as connectors, to cope with multimedia inputs from video devices, but for now the main factors have been highlighted.

10.3 Software

While most people like mobile devices to look good, the software elements of a 3G terminal will probably be more important. The software will dictate how the user will make use of, and interact with, the services and perhaps enable operators to offer personalised user interfaces that differentiate their services and networks, and foster customer loyalty.

10.3.1 Codecs

Just as with the physical aspects of 3G devices there are some fundamental software elements. Voice is an obvious use for a mobile device and it is important in the early phases of network launch that the devices are backwards compatible with existing networks in order to provide the maximum coverage possible from all current networks. The following voice codecs will be supported:

- GSM — FR (full rate);

- GSM — EFR (enhanced full rate);
- GSM — AMR (adaptive multi-rate);
- UMTS — AMR.

The AMR codecs being rolled out have the ability to adapt to the radio environment and hence, if the radio environment is good, the AMR codec will require less bandwidth for a given audio quality. This means the operator can benefit from the bandwidth it frees up and in theory at least support more simultaneous connections on the network.

As already mentioned, one of the features blown out of proportion for 3G is the ability to download video/multimedia clips. Part of the 3G standards committee is currently debating which codec should be used as a baseline implementation. A baseline H.263 has been proposed that is also compatible with MPEG4, and so it seems likely that it will be an H.263/MPEG4 combination, with the option of hardcoding the base standard into the telephone to achieve the optimum performance on a device. In fact, some companies already have single chip multimedia solutions. Even if one or both codecs are implemented this will not stop any third party from implementing their own application to be run as software on the device. The debate is still continuing and the outcome will have an impact on the mobile device and may have an effect on the network architecture as well.

10.3.2 Wireless Application Protocol (WAP)

One of the problems network operators face is that customers quite often choose their cellphone on 'look and feel' and in some cases the network operator is irrelevant. There is a real possibility that this could change in the future with new application standards being implemented. 3G standards do not define services, instead they define standards for creating services. The mobile station (terminal) application execution environment (MExE) (from ETSI/3GPP) is such a standard, defining the environment on mobile terminals for downloading and executing voice/data applications. MExE clients will have varied degrees of sophistication in capability including enhanced GUIs and will be able to negotiate via standard mechanisms with MExE servers for tailored content or service provision based on the terminal characteristics. MExE will facilitate intelligent network (e.g. CAMEL) services. As it is based on the client/server paradigm, MExE services may reside in the Internet (and interact via MExE servers). They may also be downloaded on to the MExE client as local agents or the MExE service may be installed on the client. In addition, MExE clients may communicate with one another. Since MExE will be a major change to mobile devices and it is a phased standard, it is worth taking a few moments to review the details.

10.3.2.1 MExE Classmarks

The MExE 'classmark' defines the capability of the terminal. Three classmarks (I, II and III) have been proposed and defined to date.

- Classmark I

  This is based on WAP and the terminal microbrowser (WAP telephones, even though they allow a degree of programmability via scripting, are not by definition MExE compliant, since they may not conform to the MExE security model). WAP was developed by the WAP Forum. This technology allows development of applications over wireless communications networks independent of the bearer. It is optimised for small devices and is based on the Internet client/server architecture.

- Classmark II

  This defines a specific Java virtual machine which enables various Java applications to run. Extended GUIs (e.g. browsers) and services (e.g. icons, voice recognition, soft-key customisation) may be supported.

- Classmark III

  This is under discussion (and due for imminent ratification) and is less powerful than Classmark II. It defines a specific Java configuration termed a Java 2 Microedition (J2ME) connected, limited device configuration (CLDC). The CLDC defines a restricted set of classes for devices such as mobile terminals which are resource limited. CLDC may be used in conjunction with a mobile information device (MID) profile to define a complete environment for a given class of device. MID profiles extend or modify the CLDC and define networking, timer, persistent storage, UI and application life-cycle capabilities for particular devices. Classmark III introduces MIDlets — applications that run in the CLDC/MIDP profile. The application management (e.g. downloading and registration/installation of a MIDlet application) is an additional mechanism also defined in this classmark.

Other classmarks, yet to be defined, will probably include services to further improve the GUI, such as speech recognition capabilities and enhanced interfaces, and be targeted at specific devices, such as vending machines or home audio visual (HAVI) devices.

The ability to print and store documents and so on could be handled within MExE networking capabilities (e.g. via HTTP); however, other technologies such as JINI internetworking capability may find its way into MExE classmarks.

10.3.2.2 Uses

From the user perspective, MExE-compliant devices offer personalisation — not just the look and feel but also the terminal services that can be used. The concept of the virtual home environment (VHE) — a standard 'look and feel' independent of the device type and location — is subsumed within MExE's definition.

Just as most users have unique applications running on a personal computer, so will mobile devices. Enabling this proliferation of enhanced terminal services will be the provision of services by third party developers.

There is already a sizeable community of talented Java developers, and the 'write-once, run-anywhere' capability of Java decreases time to market. Interactive games, eCommerce/mCommerce, secure banking, greater access to (full multimedia) Internet content (relative to WAP) will all be enabled by MExE standards.

From an operator viewpoint, the same benefits apply in terms of service provision. The main benefit operators have over third party suppliers in the MExE model is enhanced security. Third party service providers may well end up offering services via an operator's application service provision (ASP) centre after trial runs. It may be possible for network operators to purchase bulk mobile device orders and use a form of MExE to implement their operator/country-specific set of services and GUIs to differentiate their product from the competition. This in turn may help promote customer loyalty and reduce churn if the GUI is sufficiently attractive.

10.3.3 Other Technologies

There are many other technologies that must come together to form a fully featured 3G terminal and it is not possible to cover all the options in this brief chapter. Some of the other areas include:

- USIM/smartcard;
- security;
- operating systems;
- MMI — enabling multiple ways of accessing the device including voice recognition, touchscreen, etc;
- applications.

It is hoped, however, that the reader will begin to form an appreciation of the general area and work involved in developing 3G devices.

10.4 Conclusions

If the technologies discussed come to fruition in the next five years, some of the following should occur:

- mobile devices will be purchased in bulk as a single purchase by a central purchasing unit for all of BT's joint ventures — this yields benefits in terms of economies of scale and less need to stock numerous variants of the same type of telephone;
- it will be possible to program the devices over the air to allow a country/operator-specific user interface to be installed — this allows operators to differentiate the 'look and feel' of their services in order to help maintain customer loyalty;
- customers will also be given the option to add their own applets using the MExE standard and hence add their own personalisation to the devices;
- MExE will utilise other technologies, e.g. Bluetooth, to provide new applications such as location services and personalised information depending on that location — this will extend to the office where the device will be able to download small applets to run programs such as network synchronisation and even download printer drivers from the nearest printer;
- all screens will be colour and some will be flexible enough to be rolled up to keep the physical size small, while others will be small enough to fit into a device the size of a pen;
- with fuel cell technology, a recharge will take 10 sec and last for a month between charges;
- a common connector on the telephone will allow customers to add on multimedia devices such as recordable DVD players that can plug directly into the device and be able to transfer the contents back and forth between local memory.

Given the requirements for 3G networks and services and the current technology trends, the big question from a user's point of view may well be whether 3G will live up to the hyped-up expectations or make any difference at all. Based on current technology and the development work in progress, it is likely that there will be big improvements for mobile customers. While it is unlikely all the developments will be implemented at launch, there are enough new elements in the pipeline to make the 3G terminal a very exciting prospect.

11

THE FUTURE OF RADIO ACCESS IN 3G

J W Harris

11.1 Introduction — Current Situation

Before considering the future development of the mobile radio access networks, it is important to take stock of the current state-of-the-art. The ITU family of radio access technologies for use in the 3G spectrum is known as 'International Mobile Telecommunications 2000' (IMT-2000). Of the five technologies, three are for the paired FDD spectrum and two for the unpaired TDD spectrum (see Table 11.1).

Table 11.1 The IMT-2000 technology family.

Designation		Common name	Spectrum
IMT-DS	Direct Sequence CDMA	Wideband CDMA (WCDMA)	FDD
IMT-MC	Multi-Carrier CDMA	cdma2000	FDD
IMT-TC	Time Division CDMA	TD/CDMA	TDD
IMT-SC	Single Carrier	UWC-136	FDD
IMT-FT	Frequency Time	DECT	TDD

WCDMA and TD/CDMA, originally developed by ETSI and collectively known as UMTS, are designed as an evolution from GSM. The standards for these two technologies are now being developed by the 3rd Generation Partnership Project (3GPP), whose partners include ARIB (Japan), CWTS (China), ETSI (Europe), T1 (USA), TTA (Korea) and TTC (Japan). These are the technologies of choice for Europe, and generally any country that operates GSM networks.

cdma2000 was developed by Qualcomm and the TIA in America as a 3G evolution from the existing 2G cdma system, cdmaOne. This is now being standardised by 3GPP2, whose partners include ARIB, CWTS, TIA, TTA and TTC. This is generally considered as the technology of choice for operators of cdmaOne (IS-41) networks.

UWC-136 has been developed by the TIA as a 3G evolution from the 2G TDMA systems used in the USA, and includes certain features in common with EDGE, which is an enhancement to GSM. This is generally considered as the technology of choice for operators of TDMA networks, primarily in the 450, 800 and 1900 MHz bands in the USA.

DECT was originally developed by ETSI as a digital cordless standard. ETSI are developing this 3G-evolved system for use in the unpaired TDD spectrum.

11.1.1 WCDMA

The first release of the WCDMA specification, referred to as Release 99, was completed in December 1999. It supports a fully functional radio access network, although some of the original features have been postponed in order to freeze the specifications. Equipment designed to the Release 99 specifications will support circuit-switched voice at 8 kbit/s, using the adaptive multi-rate (AMR) codec, and, in principle, data rates up to 2 Mbit/s per cell. However, a single terminal using this data rate would consume all the resource of a cell, and would also significantly reduce the capacity of the neighbouring cells, due to the interference it would generate. In practice, operators are likely to design their networks to support up to 144 kbit/s in macro cells, with cell ranges from several kilometres to a few hundred metres (Fig 11.1). Higher data rates up to 384 kbit/s may be available over a shorter range in macro cells, and in micro cells (typically up to one or two hundred metres). This is obviously some way short of the original requirement for 2 Mbit/s, and is due primarily to the imperfect nature of the radio environment. The maximum data rates are effectively only available in an ideal deployment, with perfect radio channels, no adjacent cell or adjacent carrier interference, and perfect power control.

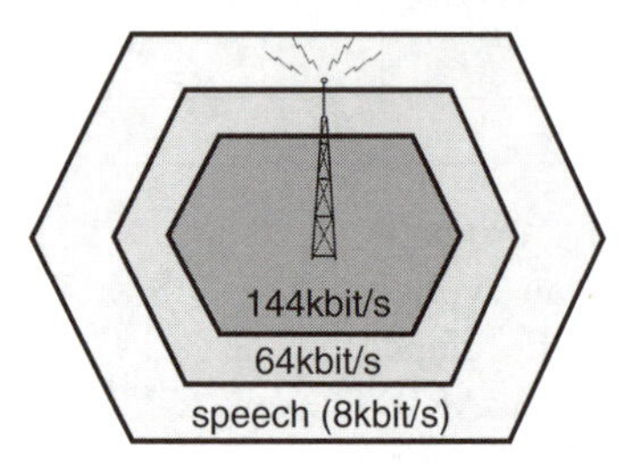

Fig 11.1 Maximum cell range for different data rates.

The use of packet-switched bearers will allow operators to approach close to the maximum capacity of the system due to the statistical multiplexing gain that packet switching delivers. The principle is simple. In circuit-switched networks, such as GSM, it is necessary to over-dimension the capacity of the network, in order to guarantee a high probability of successful access or grade of service (GoS) for a high percentage of time. Consequently, most of the time, each cell is operating significantly below its maximum capacity, as shown in Fig 11.2. Since packet data is usually tolerant of delays, the network can queue the packets and send them when there is sufficient capacity available.

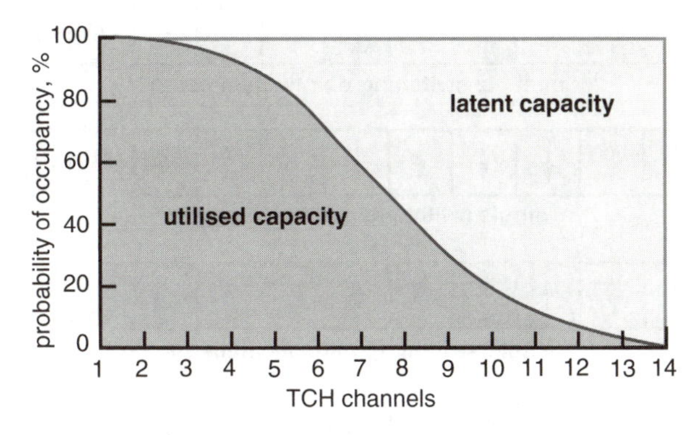

Fig 11.2 Latent capacity in 14 circuit-switched channels @ 1% GoS.

This procedure is fine for applications such as Web browsing and e-mail, which can accept that some of the packets may be delayed or have to be retransmitted. However, real-time applications such as voice, video and multimedia are particularly intolerant of delays. Thus, for real-time applications requiring high quality of service (QoS), the network is not able to make such efficient use of the latent capacity.

11.1.2 TD/CDMA

The TD/CDMA system is designed for use in the unpaired spectrum and complements W/CDMA, sharing many common parameters and integrating directly in the network architecture. It will offer the same capabilities as WCDMA, in terms of circuit- and packet-switched data rates. However, there are a number of advantages and disadvantages compared with WCDMA. The main advantage is the improved efficiency in carrying asymmetric traffic.

Being a time division duplex (TDD) technology, the data is transmitted in time-slots, of which there are 15 in a 10 ms frame. In TD/CDMA it is possible to configure each time-slot to be either up or downlink, giving asymmetry ratios from 2:13 to 14:1 (DL:UL) (see Fig 11.3). While it is possible to dynamically control the time-slot assignment according to the load, in practice there are performance implications that are likely to restrict this flexibility.

In TD/CDMA, as the data is transmitted in time-slots, the average transmitted power is less than in WCDMA, which uses continuous transmission. Since the peak power transmitted by the mobile is limited, the range of TD/CDMA cells will be significantly less than for WCDMA for low-to-medium data rates. As data rates increase, the cell sizes of the two systems become comparable.

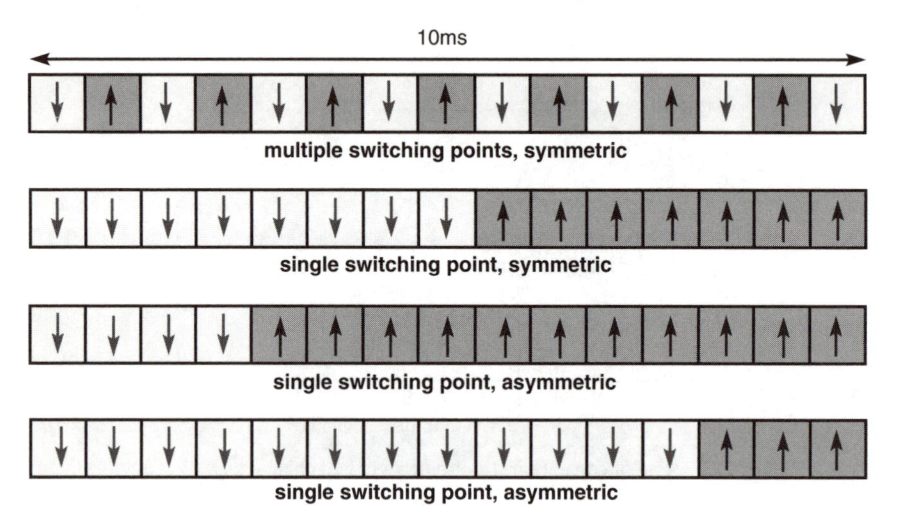

Fig 11.3 TD/CDMA time-slot configurations.

Also, in order to minimise inter-cell interference, it will be necessary to accurately synchronise adjacent cells, and to give them the same uplink/downlink configuration. It may also be necessary to co-ordinate deployment and configuration of TD/CDMA cells with the other TDD operators. Consequently TD/CDMA is likely to be used for pico cells and indoor installations providing high-bandwidth short-range coverage to offices, public buildings and shopping centres for example.

Although included in the Release 99 specification for UMTS, the development of TD/CDMA has not had the same level of support from manufacturers as WCDMA. This is because they are concentrating their efforts on the Japanese market, where there is an urgent need for WCDMA equipment to relieve the capacity problems they currently have with their 2nd generation networks. Compounding the issue is the fact that there is no TDD spectrum available in Japan, hence no requirement for TD/CDMA equipment. As a result, the equipment and terminals are not likely to be available before 2003/4.

Another potential issue is that CWTS (China) and Siemens have introduced a proposal for a narrowband TDD system into the standards. Proposed as the main 3G technology for China, TD-SCDMA has a carrier spacing of 1.6 MHz, allowing three TD-SCDMA carriers to be deployed in an unpaired 5 MHz carrier. CWTS claims that TD-SCDMA is superior in performance to TD/CDMA. However, it relies on advanced features such as beam-forming antennas and high-order modulation schemes to achieve the currently predicted performance. It is hoped that it will be possible to harmonise the two TD/CDMA technologies — yielding a single technology from the best aspects of both. However, such a process would result in further significant delays to the standardisation process, which is already some way behind that of WCDMA.

11.2 The Near-Term Future

Having established the initial capabilities of the radio access technologies for 3G, there are obvious areas with scope for improvement. From the operator's perspective, the capacity and quality of the system are key differentiators. Any enhancements that enable an operator to increase the volume of traffic that can be carried without having to invest a significant amount in additional infrastructure or spectrum are very important. Given the value that is being put on the 3G spectrum in the recent auctions, every opportunity to increase the efficiency with which the spectrum is used will be grasped by network operators, in an attempt to make a return on their investment. As far as both users and operators are concerned, increased functionality and higher data rates will be high on the agenda. Given all the initial hype surrounding 3G and the promises of such services as mobile video using 2 Mbit/s, the industry will be under pressure from the market to improve the data rates, especially as the technology quickly matures.

The next release of the 3GPP specifications for UMTS should go some way towards satisfying both operators and users in their desire for improved efficiency, functionality, quality and speed. Some of these enhancements will require software updates in the radio equipment, and as such are relatively easy to implement. Other more radical changes will require hardware to be upgraded, for both network and terminals. These are generally more expensive and take longer to implement across a complete network. The benefits will also only be available to users with the most recent terminals supporting the latest features.

Known originally as Release 2000, and now Release 4/5, the second release of the UMTS specifications address improvements in the following areas.

11.2.1 High-Speed Downlink Packet Access

There is currently an investigation into the possibility of a new downlink packet channel using a higher order modulation scheme, to support data rates up to 8 Mbit/s. Such high performance claims are of course unlikely to be achievable in practice, but the implication is that this technique will make the hitherto elusive 2 Mbit/s much more attainable. Unfortunately there is always a price to pay. In this case, the laws of physics prevail, such that the high data rates will still only be available within close range of the base-station. Also, since it relies on a new modulation scheme, existing network transceivers and terminals will have to be replaced in order to benefit from the improved performance. However, as operators move from traditional macro cells to much smaller micro cells, in order to deliver increased capacity and quality, terminals will always be relatively close to a base-station; and if the current trend for replacing a terminal roughly every 18 months continues, then providing these high speeds across the network may not be so costly after all — especially if, by increasing the capabilities, it increases the usage.

11.2.2 Repeaters

These are relatively low-cost devices that extend the coverage range of cells without providing any additional capacity. They are often located at the edge of lightly loaded cells, at tunnel entrances, underground railways, and to extend coverage into buildings. Here they receive both uplink and downlink transmissions, amplify the signals and retransmit, thus increasing the range of cells without the expense of installing a complete base-station. The challenge is that the repeater must appear transparent to the radio network, introducing minimal delay and interference, which will be particularly challenging in UMTS. They will allow operators to extend their coverage cost effectively in rural areas, or in locations where traditional base-stations are not appropriate, thus improving the quality of the network for users.

11.2.3 Terminal Power-Saving Features

The trend in GSM terminals has been to move from barely portable devices, resembling house bricks in terms of size and weight, to designer fashion items not much bigger or heavier than a matchbox. Given the market's current taste in designer hardware, terminal manufacturers are going to be under significant pressure to develop 3G terminals that not only look and feel good, but also work well (Fig 11.4). One of the biggest challenges will be delivering sufficient energy to sustain the processor-intensive WCDMA transceiver at data rates of several

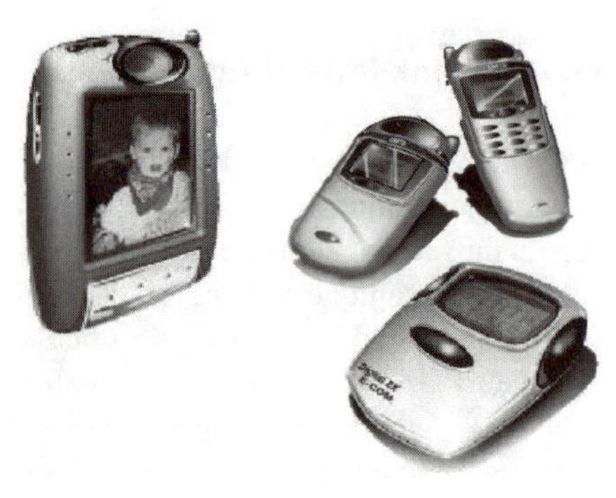

Fig 11.4 3G concept terminals.

hundred kbit/s for more than a few minutes. Techniques are being developed for minimising the power transmitted by the terminal. This will not only help to enhance battery life, but also help to reduce interference and therefore improve the capacity of the system.

11.2.4 Quality-of-Service Negotiation

One of the many features of UMTS that distinguishes it from GSM is the ability to request radio access bearers supporting a variety of quality-of-service (QoS) characteristics, in terms of data rates, error rates, delay and packet failure. The ability of the radio network to support the requested bearer QoS depends on the system load at the time of the request. In Release 99 there is no negotiation involved in this process. If the radio network cannot support the request, the request will fail. It is planned that Rel 4 will support QoS negotiation. This will reduce the bearer set-up time significantly, improving the user experience and maximising the traffic-carrying capability of the network for the operator.

11.2.5 IP Header Compression

The increasing demand for IP-based mobile services introduces some interesting challenges to the radio network, in terms of optimising the efficiency with which IP packets are transmitted. This is especially important for real-time services such as voice, video and multimedia over IP.

Future developments of UMTS will add the necessary functionality to efficiently support real-time IP traffic, covering areas such as suitable header compression and unequal error protection. The IP header provides all the information necessary to reconstruct the original data following reception. For real-time and streaming services the data contained in the header rarely changes, and as such is redundant and wasteful of valuable radio resources. With the move from IP v4 to IP v6, the contents of the header may have been streamlined, but the address fields have grown from 32 bits to 128 bits. Having almost doubled the header size, the need for effective header compression in the radio network is even more important in order to maximise the capacity of the radio access network.

11.2.6 Unequal Error Protection

In order to transmit speech as efficiently as possible in GSM and UMTS, the coded data is divided into classes of importance, in terms of maintaining intelligibility. The most important bits are more heavily protected against errors in transmission than the least important. This unequal error protection helps to minimise the radio resources required to transmit speech effectively. For data of unknown content, it is not possible to apply unequal error protection, so it must all be transmitted with the maximum protection to ensure successful transmission. This is inefficient if some parts of the data are less important than others. Identifying methods for effective unequal error protection for services other than voice will ensure optimum quality and efficiency of the radio network.

11.2.7 Smart Antennas

Smart antennas can be used to enhance the capacity and quality of mobile networks, by suppressing interference and directing the power more effectively between base-stations and terminals. There are two basic concepts for smart antennas. Switched multi-beam antennas have a number of fixed beams, with the most appropriate beam selected for each terminal. Adaptive or beam-forming antennas dynamically steer and shape the beams to optimise gain in the direction of the wanted signal, while forming nulls in the direction of interference sources. These architectures are shown in Fig 11.5.

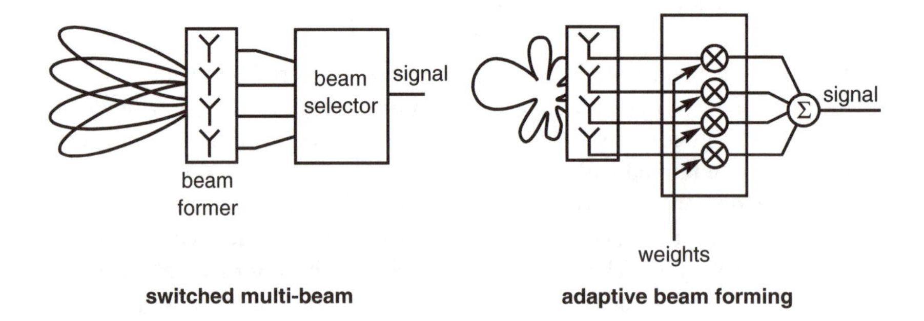

Fig 11.5 Smart antenna configurations.

Switched-beam antennas are less complex, but do not offer the same performance as adaptive antennas. The idea of smart antennas is not new, but like many such technologies, the processing requirements make the cost of the equipment uneconomical for wide-scale deployment; currently available smart antenna systems are comparable in cost to a complete base-station. However, in specific areas where interference is a problem, or where network load is increasing and there is no possibility to install additional capacity, smart antennas could provide the solution. As the technology matures and the cost decreases, future smart antenna deployment could yield capacity increases of up to 50% over traditional antenna systems. Performance gains could increase further if the adaptive antenna system is more closely integrated with the base-station.

11.2.8 IP Transport Network

The initial specification for the UMTS radio network architecture is very similar to the architecture for GSM (Fig 11.6). The base-stations (node B) are connected to their parent radio network controllers (RNC) through the Iub interface. RNCs are interconnected through the Iur, and connected to the core networks (circuit- and packet-switched) through the Iu. The transport mechanism used for these interfaces is ATM, using AAL2/AAL5. The interfaces have all been specified openly, such that equipment from different manufacturers can be mixed within the network, although operators are unlikely to adopt this practice in early networks.

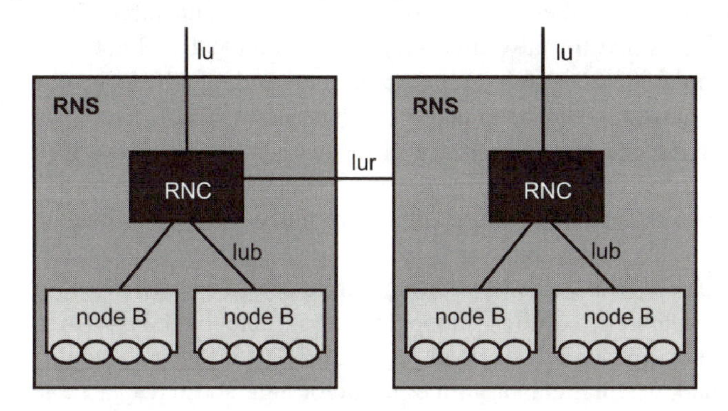

Fig 11.6 ATM-based UTRAN architecture.

There are proposals in the standards to provide the option to move from the ATM transport to an IP-based transport, as shown in Fig 11.7. There are certain advantages in terms of the flexibility that this architecture offers, but there are a number of issues that must be addressed for it to be an effective solution.

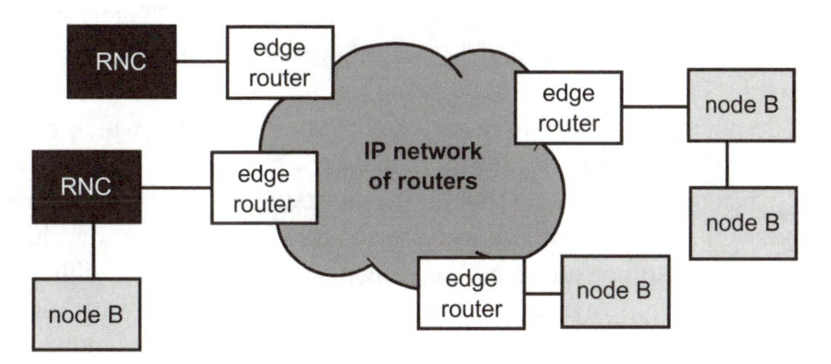

Fig 11.7 Reference architecture for IP transport network.

The mechanisms to secure the QoS parameters (service-class definition and congestion control), timing aspects (delay and delay variation) and packet loss have to be considered. Also, accurate synchronisation of the radio network nodes is required. Mechanisms for minimising delay variation and clock-frequency differences between nodes must be provided. While the move to an all-IP communications world continues, the benefits of IP transport over ATM/AAL2 in UMTS are not clear, and will vary between different operators, depending on existing transport network infrastructure, other applications using the same transport, available physical links, etc.

11.3 The Future for TDD

As mentioned earlier, the technology designed for the unpaired TDD spectrum is not as mature as that for the paired FDD, and equipment will not be commercially available until 2003/2004. While there is no outstanding advantage over WCDMA, TD/CDMA has the potential to deliver higher bandwidths for services with a high degree of asymmetry more efficiently, albeit in much smaller cells. Thus for the four operators in the UK for whom the spectrum package they bought included 5 MHz of TDD, there exists a potential commercial advantage.

Typically TD/CDMA might be used as the access medium supporting managed networks for corporate clients, offering wireless office solutions. It could also be deployed in conference centres, shopping centres, hotel complexes, airports, railway stations, motorway service areas, or any location offering sufficient isolation from the FDD band, where highly asymmetric and/or high-speed data access is required.

Being aimed primarily at hotspot rather than wide-area coverage makes it unlikely that TD/CDMA-only terminals will meet people's communications needs. Since initial UMTS terminals will be dual-mode WCDMA and GSM, terminals supporting TD/CDMA will probably be tri-mode (WCDMA, TD/CDMA and GSM). Unfortunately it will not be possible to have simultaneous working on mixed technologies in a multi-mode terminal. So, for example, it will not be possible to have a voice call on WCDMA and a high-speed data call on TD/CDMA from the same terminal — the reason being that the high-power FDD uplink transmission will saturate the TDD receiver.

One of the most frequently discussed applications for TDD, which exploits the efficient support of asymmetric traffic, is that of a high-speed bulk up/download. The analogy has been drawn with mobile 'filling stations', whereby TDD pico cells are deployed in locations where users can conveniently use the high-speed capabilities of TDD to perform a bulk up/download. For example, with TDD cells distributed around a railway station, in particular along the platforms, users could use the time while waiting for a train to synchronise their e-mail, download Web pages, music or video for off-line browsing.

This system would also serve the passengers on the trains passing through the station, who would also have access to the highest possible data rates while stationary at the platform, rather than having to cope with the relatively low rates available from the wide-area FDD cells.

TD/CDMA would also be suited to the provision of wireless LAN-type access for corporate networks, with the advantage that they could be connected to the wide area network and so provide ubiquitous coverage and seamless service, handing over from the corporate network to the national network. This would of course require a significant investment in radio equipment for customer's premises, and would have to offer significant benefits in terms of mobility and integration to the wide area cellular network over fixed network infrastructure or wireless LAN systems.

The key points in favour of TD/CDMA are that it:

- provides additional capacity, assuming terminal support;
- is potentially capable of supporting higher data rates;
- offers more efficient support of asymmetric traffic;
- allows guaranteed QoS, being licensed spectrum;
- allows ubiquitous mobility, with seamless connection to the wide area network.

11.4 Terminals and the Role of Bluetooth

If we consider the types of services and applications that 3G terminals will be supporting, it is evident that the future terminal paradigm is likely to be somewhat different to today's voice-centric devices. It is also clear that low-power, hand-held terminals, with limited displays, processing power and storage, are unlikely to meet the needs of power-users wanting to exploit the higher data rates supported by UMTS. PDAs and laptops are likely to remain the main computing platform for a significant segment of this mobile market — in order to meet their communications requirements the model shown in Fig 11.8 could be envisaged.

The mobile user will still want a small voice terminal, including a limited graphical display for low-to-medium rate data applications, such as WAP browsers and e-mail access. These would generally be low-power WCDMA/GSM transceivers with Bluetooth connectivity.

The Bluetooth link would provide connectivity to other devices, including a remote headset, PDA or laptop, in the same way as exists today with cables or infra-red.

Bluetooth is a low-power, short-range radio system that has been developed by a consortium of manufacturers to address the issue of wireless connectivity for computing and telecommunications devices. It operates in an unlicensed band of spectrum at 2.4 GHz, and offers data rates up to 1 Mbit/s over a range of 10 m.

As an optional facility the user could purchase a PC card 'radio', encapsulating a UMTS transceiver. This would support higher data rates than the voice terminal, due to the increased power available from the host device. This card would also include Bluetooth connectivity, allowing it to link with the voice terminal. In this

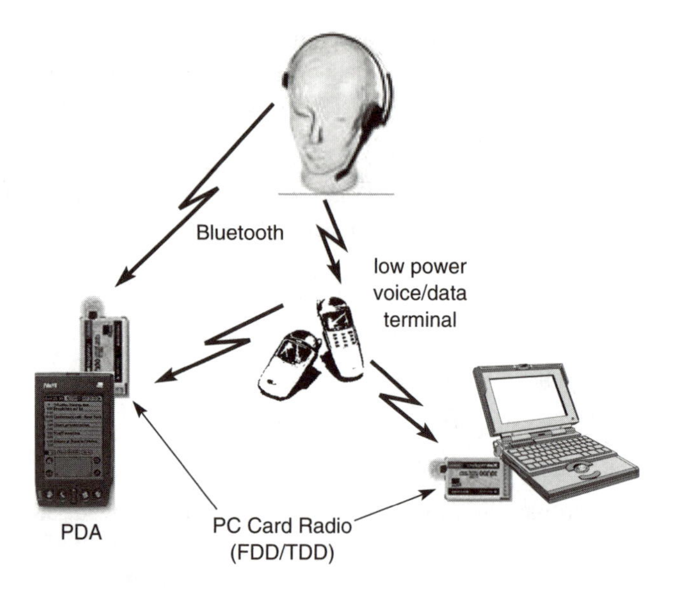

Fig 11.8 Terminal connectivity model.

mode the PC card would provide the air interface, supporting, for example, high-speed data, together with a voice connection through the Bluetooth link to the voice terminal.

By developing a model for terminals such as this, it could be envisaged that the cost of the 'radio' in the PC card would be relatively low. Assuming this to be the case, users would be more likely to upgrade the radio card for additional functionality or future enhancements offering higher data rates. This would also allow a relatively inexpensive solution to the problem of multi-mode terminals to support global roaming.

11.5 Further Ahead

Most of the technologies that have been discussed so far are likely to be available commercially in the next few years. There are other techniques that have been developed to the concept stage, but are not yet available due to technology limitations — one of which is software defined radio (SDR).

SDR uses digital-to-analogue conversion and advanced digital signal processing to synthesise radio signals so that the air interface of a mobile network can be controlled and programmed using software. This would allow the frequency band, channel bandwidth, modulation and coding schemes of the transceivers to be dynamically reconfigured in software, rather than requiring either separate devices or complex multi-mode terminals (Fig 11.9).

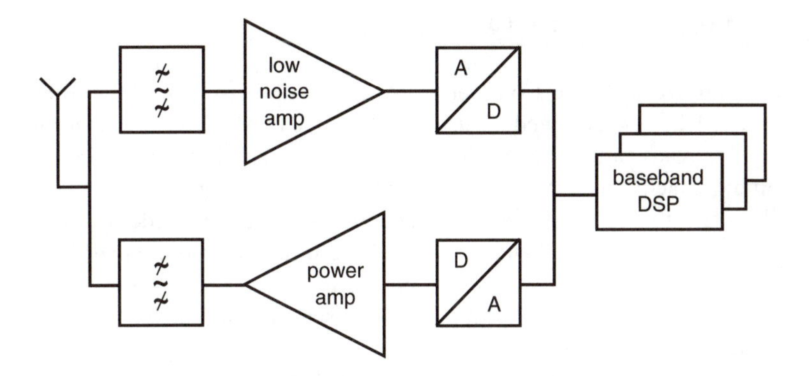

Fig 11.9 Ideal software defined radio transceiver.

Unfortunately the current state-of-the-art for digital-to-analogue converters and baseband processors falls some way short of that required for a full implementation of SDR. What is likely to emerge is an evolutionary approach to SDR, with the interface between dedicated and programmable hardware moving towards the antenna as the technology advances. Since sampling at RF is not currently possible, the radio signal has to be converted to an intermediate frequency, with programmable up- and down-converters performing the filtering, modulation and demodulation.

SDR could potentially be used for both base-station and mobile terminals, although the current emphasis is for base-station deployment; terminals are unlikely to have sufficient processing power or battery life to support SDR for some time to come.

While there are obvious advantages for SDR terminals supporting multiple radio systems, the advantages for network operators deploying SDR base-station equipment include:

- single platform supporting multiple air interfaces;
- migration and upgrade by software updates;
- flexible network roll-out;
- dynamic allocation of carriers, bandwidth and modulation schemes.

All of these benefits equate to potential reduction of network operating costs, and enhanced service propositions.

Although SDR is not commercially available at present, it has significant potential and it is very important for operators to keep up to date with developments so that they are in a position to exploit this new technology when it does become available.

11.6 Future Spectrum Availability

At the recent World Radio Conference (WARC 2000), a block of spectrum from 2500-2690 MHz was identified for use by IMT-2000 (UMTS) as an extension to the core band already allocated at WARC 92. It is very early days in terms of determining exactly how this spectrum will be used. As anticipated, it is proving difficult to use this spectrum in a frequency division duplex mode, given that the duplex distance in the core band is 190 MHz, and that the requirement for duplex separation would effectively waste a significant portion of this new spectrum. It would seem likely therefore that it would be used in TDD mode, using the most appropriate TDD technology.

Another possibility would be to use it as an FDD downlink-only carrier. Support for variable duplex distance in WCDMA could be taken to an extreme, such that multiple downlink carriers are assigned to a single uplink. This would allow WCDMA to support the predicted highly asymmetric traffic much more efficiently than is currently possible. It was also agreed that the existing 2nd generation bands (900 MHz and 1800 MHz in Europe) be designated for UMTS use in the future, potentially allowing existing operators to migrate their existing spectrum to 3G.

However this additional spectrum is used, it is clear that in order to fully exploit its traffic-carrying capacity, new terminals will have to be introduced to replace those supporting only the core bands.

11.7 Conclusions

As the operators and mobile users prepare for the launch of the first UMTS networks, the industry is already preparing to improve the capabilities of the proposed systems. These enhancements will manifest themselves in various forms, including increased data rates, capacity, quality of service, functionality, etc. This chapter has introduced some of these developments, some of which are just around the corner, others beyond the horizon. Some may fall by the wayside, others are still waiting to be developed. What is important is that there is still room for improvement. As our dependence on mobile communications grows, the demands on the systems will continue to increase, and what seem like unnecessary or unachievable capabilities today will be taken for granted in a few years' time.

12

EDGE MOBILITY ARCHITECTURE — ROUTING AND HAND-OFF

A W O'Neill and G Tsirtsis

12.1 Introduction

Telecommunications networks are rapidly progressing towards an all-IP architecture. This trend is not restricted to fixed networks. Second generation cellular networks have been modified to provide limited data services, and third generation systems currently undergoing roll-out have been designed to deliver IP connectivity to the end user [1, 2]. Projecting technological trends, it is seen as inevitable that the future global telecommunications network will consist of a routed/switched IP core (much of which will be optical) accessed via a wide range of edge technologies. Many of these edge technologies will support mobility based on the continuing advancements in wireless technology. Internet service providers will increasingly want to support both fixed and mobile users. With IP routing technology being pushed out to the network's edge, it becomes less cost effective to support the various modes of layer-2 edge mobility management that accompany today's cellular and PCS technologies. Rather, unified solutions become advantageous to domain operators, wherein mobility management becomes an integral component of an IP layer routing protocol.

Internet routing protocols have been traditionally designed from an assumption that the location of an IP interface in the topology is static. In addition, they assume that address allocation within the topology will aim to provide multiple levels of IP address aggregation such that routing protocols can deal with address prefixes, rather than large numbers of host routes. Within this framework, traditional intra-domain protocols, such as OSPF, need only react to infrequent changes to the network due to link or router failures or permanent modifications to the addressing scheme or the topology. In contrast, mobile *ad hoc* network (MANET) routing protocols have been developed to address situations whereby the mobile nodes have permanent IP addresses but can still rapidly roam through an *ad hoc* topology, leading to the need for alternative routing technology and the general loss of aggregation opportunities.

This chapter considers a third family of routing protocols, for the case in which the core topology is essentially fixed but where the end systems may be mobile. This is a mixture of the traditional prefix-routed scenario of the fixed Internet and the classical edge mobility scenario that is today supported by cellular networks, primarily as part of the cellular technology elements (GSM, GPRS, etc). Migrating the latter mobile routing functionality to layer 3 — to release all the end-to-end internetworking benefits which have aided the deployment of the Internet — would tend to suggest a fusion of the MANET and traditional routing protocol architectures. The primary aim is to move the IP interface location in the routing topology as the mobile changes base-station (BS) so that active IP sessions are maintained.

This form of routing is referred to as mobile enhanced routing (MER) within an edge mobility architecture [3].

These networks can be considered to have a single IP routing protocol that runs between routers in the EMA domain. Some of these routers may be traditional access routers (ARs) acting as the first IP node for collections of fixed and roaming hosts (e.g. PSTN, cable head-ends, corporate access, GPRS gateways). Others may be future access routers collocated with, or serving as an aggregator for, base-stations equipped with a (potential) diversity of wireless technologies such as CDMA, TDMA, radio LANs, etc. The radio layers are assumed to provide the well-known layer-2 hand-off models and other capabilities including break-before-make, make-before-break, power measurement, mobile-assisted hand-off and security features. To facilitate internetworking, inter-access router co-ordination is assumed to use IP-based communication using messages which are abstractions of the messages which are today carried in cellular technology-specific messages, often via central processing elements.

The remainder of this chapter is structured as follows. Section 12.2 describes an edge mobility architecture (EMA) for the generic support of edge mobility, with the aim of being general enough to support a range of different routing protocols, as well as enabling hand-off between diverse types of cellular technology through capability exchange between radio-equipped ARs. It will become clear that the routing approach put forth here needs to be coupled with a companion paging architecture for location management, but this is not covered in this chapter. In section 12.3, a detailed description of the proposed approach for mobile enhanced routing is provided. In section 12.4, the relationship between and convergence of EMA and mobile IP (MIP) is described and in section 12.5 the scalability benefits of the EMA:MER approach is discussed. Detailed simulation performance numbers are not detailed in this chapter but are included in referenced documents. Finally, some concluding remarks are given in section 12.6.

12.2 Edge Mobility Architecture

The proposed architecture assumes that modifications to either MANET or traditional routing protocols are possible which will enable these protocols to comply with this architecture and hence facilitate a message set and control model which has a degree of protocol independence. The architecture has six main components, the first being the use of mobile IP across provider boundaries to facilitate the temporary movement of an IP address (on a mobile terminal interface) away from its home domain while maintaining active sessions. This bounds the immediate discussions to intra-domain routing issues. The other five components are described below and shown in Fig 12.1:

- the provision of a modified intra-domain routing protocol which provides prefix-based routing within a domain, with each prefix representing a block of IP addresses allocated to each AR in the domain, as well as host routes to support mobile host migration away from the allocating (IP address) AR;
- the provision and use of virtual links for routing exchange and messaging between co-operating ARs to exchange capabilities, and to effectively and

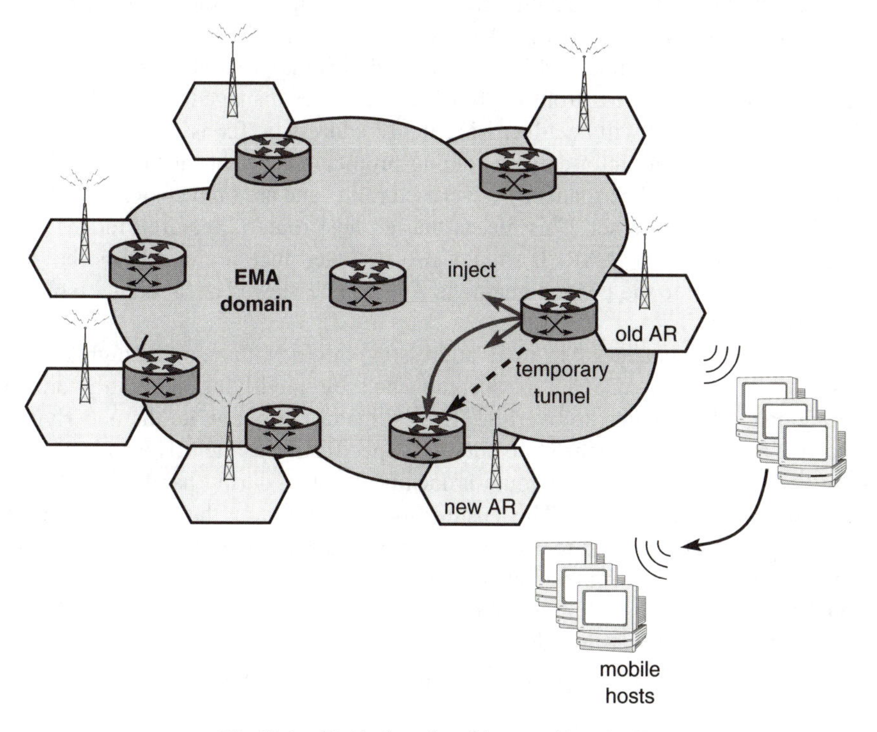

Fig 12.1 EMA domain with routed hand-off.

locally manage the hand-off of the responsibility for, and routing of, the mobile terminal and its associated IP address;

- the provision and use of a temporary tunnel to redirect packets in flight from the old AR towards the new AR while routing converges;
- the ability to inject a host route for the mobile;
- a method to return the allocated IP address to the allocating AR on mobile session termination at a different AR in the same domain.

The reasons for each of these components will be explained in the following subsections which give examples for CDMA (i.e. make-before-break) and TDMA (i.e. break-before-make) layer-2 hand-off. Using an appropriate set of hand-off messages over IP ensures that a wide range of radio-technology-specific hand-off models can be accommodated within a single IP model, to allow for internetworking of IP over those diverse technologies.

12.2.1 Mobile Session Start-up

When the mobile either has data to send or has been paged for incoming traffic, the mobile connects to the nearest/best AR and is brought into the IP routing domain. It achieves this by requesting and being allocated (with appropriate AAA checks) a temporary local IP address from the block of addresses managed by that AR. This allocating AR (AAR) will be advertising the IP address prefix associated with that address block into the intra-domain routing protocol such that ‘at home’ mobiles have a proactively and permanently advertised route, and are immediately reachable by all hosts in the Internet. This means that no host route is required for the MH while it remains at the AAR. It should also be noted that any end hosts that are statically attached to the EMA domain via ARs can be viewed as ‘at home’ mobiles who never move.

When a mobile changes AR, its IP address(es) moves with it so that higher-layer sessions are unaffected. This is accomplished by modifying the intra-domain routing, using host routes, to overrule (longest match) or overwrite the underlying, proactive prefix routing to the allocating AR. Specifically, as the MH wanders away from the AAR, a host redirect route is locally injected, during hand-off, between geographically adjacent ARs. This ensures that any traffic on the aggregated AAR route is ‘peeled off’ and redirected to the new AR. Subsequent movement results in additional host redirect routes that progressively ‘peel’ the incoming traffic away from both the prefix route for the AAR, and the previous host redirect route. Hence, the further we wander at the edge, the further the sequence of host redirect routes will extend the redirection away from the aggregated AAR prefix route (and associated AR).

12.2.2 IP Session Dynamics

When the mobile is inactive for short periods during an IP session, the radio layer has mechanisms and timers which enable the radio resources to be released to other users. An 'IP session timer' is used in the AR and MH to identify sufficient inactivity (which could be application, terminal, user or service class (tariff) specific) to terminate the dynamic IP session, so releasing the associated radio resources, temporary host address and associated routing state. This enables high multiplexing of the temporary IP address space, and better utilisation of routing and radio system resources. For mobile users who wish to be able to maintain the temporary IP address for long periods (e.g. for application, business, usage pattern reasons), the IP session timer can be set very long (with potential tariff implications). As this could result in long-term host-specific IP routing state in the system which may be undesirable, an additional 'IP routing timer' is used in the MH and AR which causes the host routing state to be removed while keeping the IP address. This is clearly a per-user policy decision as to the appropriate value for this timer, but a large default value can also act as a safety mechanism.

12.2.3 Mobile Node States

Summarised below is the IP level view of the state of a mobile in EMA that may be mapped to existing cellular states. Specific local terminology is used here due to confusion in naming between the various global standards in existence, and because the focus here is on the various inactivity timers whose lengths are related to the users' profiles/privileges, and whose expiry moves the mobile between specific states.

- EMA active

  The mobile is presently sending and receiving IP data traffic. The MH has a local IP address and has a host-specific route pointing to it in the EMA domain. The radio layer (L2) and IP layer (L3) are obviously up and movement between access routers generates an EMA IP hand-off.

- EMA hot standby

  The MH moves to hot standby when its L2 inactivity timer (not sending/receiving data) expires. This takes the L2 down temporarily while the L3 is still up, which releases the radio layer resources to other users. The MH therefore maintains the current IP address, and has an EMA host route for that address in the EMA domain, while movement between access routers generates an EMA IP hand-off. This feature further improves the utilisation of the radio layer by decoupling the IP address and layer 3 resources from the L2 radio resources.

- EMA warm standby

  On expiry of the IP routing timer due to prolonged IP data inactivity, the network can flush out the EMA host redirect route for the MH but still keep the current local IP address (in session) in preparation for subsequent IP data. The L2 is down, and movement results in location updates either due to paging location area (LA) changes (movement-based) or when the location update timer expires (timer-based). Location updates are sent to the AAR and to the global paging system. This feature is useful for users who want to be able to keep the temporary local IP address for business/application reasons, but who are actually only intermittent IP data users. The burden of such users can be reduced by removing their host routes on inactivity, and then re-injecting a host route back to/from the allocating access router when they wish to send, or when data is received at the AAR for that IP address.

- EMA cold standby

  On expiry of the IP session timer, the session IP address is returned to the AAR and any EMA host redirect state is flushed. The MH now optionally only has its home address from its home link. The L2 is down and subsequent movement generates location updates only to the global paging system, because the MH now has no AAR, and local fast route set-up is not required. This feature is designed to avoid hoarding of IP addresses when the user is inactive, so that the address can be returned to the AAR for another user.

- EMA off

  The MH is switched off and is neither sending location updates nor pageable.

12.2.4 EMA Hand-off

A hand-off occurs when the MH is in either the EMA active or hot standby states, and it needs to use a different first-hop access router. The result of hand-off is that the route pointing towards the EMA IP address of the MH is moved between access router(s). During hand-off [4], the access routers conspire to try to achieve the impression of layer 3 'make-before-break' hand-off (seamless) even for 'break-before-make' radio systems through the use of inter-access router tunnelling, user protocol state transfer [5] and buffering. The hand-off can be initiated either by an MH or the network, and can be rooted (i.e. initiated) at either the old or new access router. The old access router (OAR) is used to co-ordinate a forward hand-off when the new access router (NAR) is known in advance as a result of either network or MH-based movement prediction and associated performance measurements. The NAR is used to co-ordinate a reverse hand-off when the NAR is not known in advance by the MH or the network, as a result of either network failure (e.g. old radio link lost), insufficient network intelligence (e.g. no inter-technology hand-off signalling), or unpredictable user behaviour (e.g. swapping PCMCIA network cards).

12.2.5 Mobile Session Completion

It is clear that the migration of IP addresses away from the allocating AR can lead to address exhaustion (in IPv4) and a gradual degradation over time of the usefulness of the proactively advertised AR address block prefix. It is therefore critical that, at the moment that the mobile finishes active sessions at a distant AR, the IP address is returned to the home AAR (in IPv4). This can be modelled as a virtual mobile moving from the distant AR back to the home AR and then locally returning the IP address. This can be accomplished using similar mechanisms which are used to support real inter-AR hand-offs, with the ARs acting as proxies for the virtual mobile. Their aim is to co-ordinate the removal of all host-specific routing entries in the domain as a result of previous mobility away from the home AR while returning the address.

12.3 Mobile Enhanced Routing (MER)

The preceding architecture does not specify how an MER algorithm creates or modifies its host and prefix-specific IP forwarding entries, and various approaches are possible. The problem of MER is divided into two sub-problems — inter-AR routing and host-specific routing. The inter-AR routing should be continuously maintained, i.e. proactively, whereas host-specific routing should be maintained only as needed, i.e. on-demand or reactively, for scalability reasons.

Inter-AR routing is prefix-based, i.e. each AR advertises a prefix address into the EMA domain covering a block of host addresses that it controls. Each host is allocated an interface address covered by the allocating AR network prefix. While the host is 'at home', packets are routed to the host via this network prefix. This is undertaken so that:

- packets can be efficiently routed to both fixed and 'at home' mobile hosts;
- a virtual, bi-directional link exists between any pair of co-operating ARs for their communication during mobile hand-off.

Host-specific routing is required only when the mobile host moves away from its allocating AR. Host-specific state is injected in the network during hand-off to overrule (via longest prefix match forwarding) a mobile host's 'at home' prefix-based route, which redirects packets to that mobile host's current AR location.

With the objective of building a large-scale MER domain, the temporally ordered routing algorithm (TORA) appears potentially well-suited for use as an EMA routing algorithm [6]. TORA was originally conceived as a MANET routing algorithm where it is intended for use in large-scale, dynamic, bandwidth-constrained MANETs. The principal objective behind its design is the achievement of 'flat scalability', i.e. achieving a high degree of scalability (measured as the number of routers in a domain) with a 'flat', non-hierarchical routing algorithm. In its operation the algorithm attempts to suppress, to the greatest extent possible, the generation of far-reaching control message propagation. With TORA, such suppression may or may not be feasible depending on the topology. As will be shown, a key to achieving highly scaled, robust EMA routing with TORA turns out to be an issue of topology design.

TORA supports loop-free, multipath routing realised as a consequence of the usage of totally ordered 'heights'. The provision of multipath routing makes the protocol amenable for load sharing and traffic engineering. The algorithm also has the potential to support fast restore via its link reversal mechanism[1] based on the availability of fine-grained link status sensing (possibly from layer 2).

12.3.1 TORA Concepts

TORA was originally specified as an on-demand routing algorithm, but this mode of operation is not generally required and a more proactive mode has been specified. Because TORA proceeds independently for each destination, it may operate proactively for certain destinations and reactively for others. In the proposed EMA context, separate instances of TORA:MER will proceed proactively for each AR and proceed reactively for each mobile host in an edge mobility-enhanced mode as necessary.

TORA operates with respect to 'nodes'. Each node is assumed to have a unique node ID (NID). A NID is a polymorphic identifier, and may be either a router ID (RID) or a destination ID (DID) depending on the context. In a manner similar to OSPF, TORA uses a router ID to uniquely identify a TORA router separately from its interfaces. A destination identifier in TORA is a destination network prefix, composed of an interface IP address and a network mask. Consequently, the neighbour set N_i that lists a node's neighbours by NID may actually contain two different identifiers. A neighbour may be identified as a router (by its RID) or as a destination (by its network prefix) or frequently as both, with multiple entries in the neighbour set table. For the subsequent discussion, it is assumed each node i is always aware of its neighbours in the set N_i.

TORA proactively and/or reactively builds, and then maintains, a directed acyclic graph (DAG) rooted at a destination. For a given destination, each participating node i is assigned a height defined as an ordered quintuple $H_i = (\tau_i, oid_i, r_i, d_i, i)$. No two nodes may have the same height (i.e. the set of heights is

[1] Due to space constraints a complete description cannot be given here of the protocol's link reversal operation in response to arbitrary link failures; see Park and Corson [6, 7] and Park [8] for these details. This chapter will only touch on aspects of the link failure processing as necessary.

totally ordered). Height comparisons are performed lexicographically. Starting with the τ_i value, comparison tests are for 'less than' or 'greater than', with equality resulting in the comparison proceeding element-wise towards the final element, the NID *i*. Information may flow from nodes with higher heights to nodes with lower heights over 'directed' links. Each link is assumed to allow two-way communication (i.e. nodes connected by a link can communicate with each other in either direction). Each initially undirected link may subsequently be assigned one of three states — undirected, directed from node *i* to node *j*, or directed from node *j* to node *i*. If a link is directed from node *i* to node *j*, node *i* is said to be 'upstream' from node *j* while node *j* is said to be 'downstream' from node *i*. Conceptually, information can be thought of as a fluid that may only flow downhill over downstream links (see Fig 12.2). By maintaining a set of totally ordered heights at all times, it is easy to see how loop-free, multipath routing is assured[2]. Information would have to flow uphill to form a loop, and this is not permitted. Due to the mobility of some nodes, the set of active links in the system is changing with time (i.e. new links can be established and existing links can be severed).

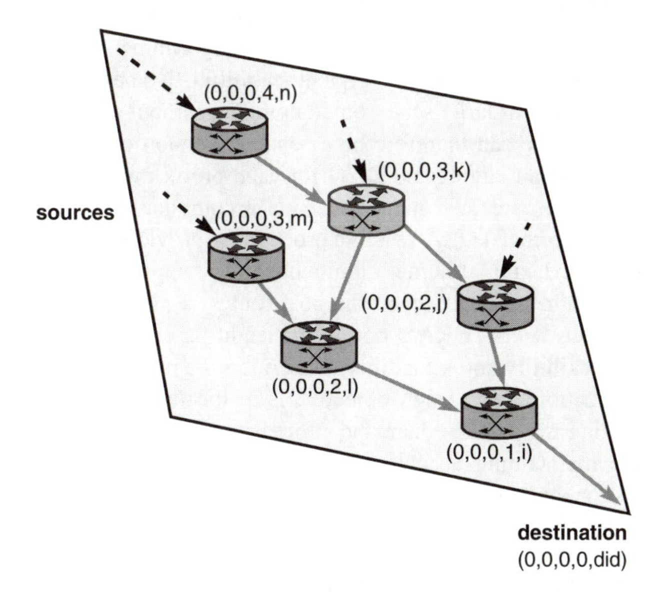

Fig 12.2 Downhill flow of information along DAG.

Conceptually, the height quintuple associated with each node is combined into two parameters — a reference level, and a delta with respect to the reference level. The reference level is represented by the first three values in the quintuple (a triple),

[2] As with many on-source-routed protocols such as OSPF, transient loops may exist in the tables between adjacent routers that are exchanging routing control messages. With TORA these loops are only of the 'ping pong' type, i.e. between adjacent routers, and are very short-lived in wired networks due to the speed and reliability of neighbour information exchange.

while the delta is represented by the last two values (a double). The first value in the reference level, *i*, has three meanings. If equal to zero, it indicates that the height value has remained 'unchanged' since the DAG was initially constructed or was last optimised (this is the state of all heights in Fig 12.2). If positive, it is a time tag representing the 'time' of a link failure somewhere in the network. If negative, it represents a route 'freshness' value (the more negative, the fresher the route) generated in response to hand-off-induced mobility.

Much of TORA's original protocol mechanism deals with reaction to link and node failures. Many of these details are not relevant to the discussion here, and the reader is referred to Park [8] and Park and Corson [6, 7] for this information. Here the focus is on the aspects of TORA necessary to understand its operation as an MER algorithm within the EMA. Under appropriate topological conditions, TORA's reaction to link additions and failures can be highly localised. This is a key property which is exploited based on the realisation that, viewed abstractly, the 'make' and 'break' operations in cellular networks correspond to link 'additions' and 'failures', respectively, in a unified mobile host/fixed router network. The subsequent lack of large amounts of potentially far-reaching control message propagation — a feature common to shortest-path algorithms — afford TORA its relatively quick convergence and consequent stability [9]. These properties appear desirable for the design of large-scale routed domains without any consideration for mobility support. There can therefore be a separate version of TORA, running on each AR, advertising an aggregated DAG for each prefix owned by that AR. The AR aggregated DAGs are then maintained via a combination of proactive route advertisement and normal TORA reactive processing, providing large-scale routing support for both fixed, and 'at home' mobile hosts. To support the movement of the mobile hosts through the injection of the host routes, a mechanism is sought that operates in harmony with TORA's notion of height-based routing, and permits a large degree of flexibility and scalability concerning the method and scope of host-specific state injection. The design objective is to localise the scope of hand-off-induced messaging so as to reduce the processing load on routers as much as possible, while maintaining acceptable routing efficiency. Domain scalability is clearly the end goal and the resulting mechanisms are outlined below.

12.3.2 TORA Hand-off Processing

TORA:MER differentiates nodes into two classes — routers and hosts. Routers execute the full MER protocol while hosts execute only a limited state machine that does not involve packet forwarding. Base-stations are routers (ARs), and mobile hosts (MHs) act as the hand-off between routers. In general, routers may also be mobile (e.g. mobile *ad hoc* networks), but that case will not be considered here.

The TORA:MER protocol operates reactively in response to controlled topology changes (i.e. hand-off) and uncontrolled topology changes (i.e. failures) to build and maintain host routes. How TORA:MER responds to controlled topological changes, whether foreseen or not, is described below.

In certain wireless technologies (e.g. GSM), hand-offs can be predicted based on signal-to-interference measurements at nearby BSs. After the appropriate hand-off criteria are reached, a hand-off procedure can be initiated from an OAR to an NAR in response to a predicted topological change in a controlled fashion. In other instances, e.g. with technologies not supporting hand-off prediction, the unpredicted loss of a link should not be immediately interpreted at the OAR as an undesirable link failure. Instead, the OAR should wait for a time to see if the MH reappears, connected either to itself or to an NAR. If it appears at an NAR, the OAR again treats this as a controlled topological change and reacts accordingly. If it does not appear, the OAR eventually declares link failure and reacts in accordance with an uncontrolled change.

A general mechanism is now given for moving a destination identifier from one node to another in a TORA domain, which is then specified for the case of hand-off processing.

12.3.2.1 Transference of Destination Identifiers

TORA nodes may be associated with multiple destination identifiers, and a separate DAG may be built for each identifier. It is possible to change the association of a destination identifier from one node to another while preserving routing integrity by lowering the destination's height. For example, as shown in Fig 12.3, when a DAG is initially constructed for a destination identifier, it is assumed to have a **zero** height (0, 0, 0, 0, did). To transfer the destination, its height is lowered to (–1, 0, 0, 0, did). In essence a new negative reference level (–1, 0, 0) is defined for the destination. A temporary virtual link (shown as a dashed arrow) to the new node (node *k*) is built at the old node (node *h*) so that the DAG remains well defined with only a single destination. Otherwise, the TORA algorithm at the old node would have to react since it would have lost its last downstream link to the destination. Note that the routing is still loop-free. However, for the old node *j* to forward packets to the destination identifier, it has, in effect, to tunnel them to the new node *k*, which would then forward them on to the destination. The new node *k* accepting the destination assumes a height (–1, 0, 0, 1, k), indicating that it is one hop from (and still higher than) the destination.

What remains now is to remove the virtual link. This can be accomplished by sending a unicast-directed update (UDU) message to the old node *h* redirecting a portion of the DAG. In the example, node *k* generates a UDU and sends it towards node *h* via node *j*. The UDU carries a copy of the destination identifier, the transmitting node's height for the destination identifier, (–1, 0, 0, 1, *k*), a destination identifier for the old node (not shown), and a designated next-hop receiver (*j*). Note that a copy of the UDU should also be sent to any other neighbour of node *k* that

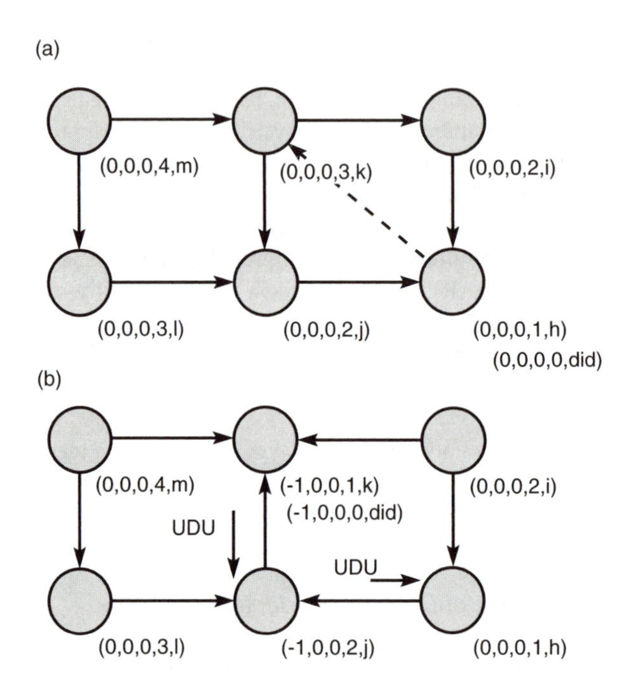

Fig 12.3 Moving the destination of a DAG.

already knows its height (nodes *i* and *m* in the example) to inform them of the height change. However, only node *j*, being the designated receiver of the UDU, may continue forwarding it towards the old node. The designated receiver *j* adjusts its height by adopting the reference level of the UDU and increasing its δ value by one before forwarding. It also picks another designated receiver, node *h* itself in this case. On receipt of the UDU, the old node will remove the virtual link. This process can be repeated as the identifier is moved from node to node, with the δ value being decreased by one each time.

The preceding example described the movement of a host interface identifier associated with a router — in effect, the movement of an attached host from one router to another. In a similar fashion, it is also possible to move a destination identifier (a network prefix) associated with one router's interface to another router's interface (this permits subnet mobility). The visibility of the destination identifier's new location can be increased by more broadly advertising the new location on hand-off. In the limit, one could advertise a new reference level throughout the entire domain, effectively performing a re-optimisation of the DAG for the most negative reference level. There is a trade-off between routing efficiency and advertisement scope, and the trade-off is topology dependent.

12.3.2.2 EMA Hand-off Messaging

If MH movement is predicted, then the OAR may be informed by the MH (if mobile assisted operation is implemented) with a host tunnel initiation (H-TIN) packet to the NAR as shown in Fig 12.4 [4]. This causes the OAR to build a temporary, soft-state tunnel towards the NAR and to send a tunnel initiation (TIN) packet to the NAR. This message may give the NAR advance warning of hand-off. The NAR can also forward a hand-off hint (HH) to the mobile to indicate to it that the NAR is ready to receive it. The tunnel can serve to help avoid packet loss during any link dead-time. This sequence of events, the tunnel's construction and the HH message are all optional. What is not optional is the construction of a virtual link at the OAR. If hand-off is predicted, this virtual link is accompanied by a tunnel and is terminated at the NAR. If hand-off is not predicted and the link to the MH is suddenly lost, a virtual link to the MH itself is retained for some time while the OAR awaits notification of the MH's location.

Otherwise, the EMA hand-off model has its focus at the NAR and all mandatory messaging begins there. On arrival at the NAR, the MH (operating in mobile-assist mode) brings up a new link for IP purposes (i.e. make) by sending a host hand-off request (H-HR) message to the NAR. This triggers the NAR to send a hand-off request (HR) message to the OAR. The OAR responds with either a hand-off initiation (HI) (i.e. AAA information and associated state) packet if hand-off is permitted, or else with a hand-off denial (HD) packet. The HR packet may be repeatedly sent until either an HI or HD is received, or it is determined that the OAR is unreachable. If hand-off is permitted, the HI packet begins a three-way handshake to transfer control of the mobile to the NAR. On receipt of the HI, the NAR initiates routing redirection by sending a UDU towards the OAR. This is sent reliably hop-

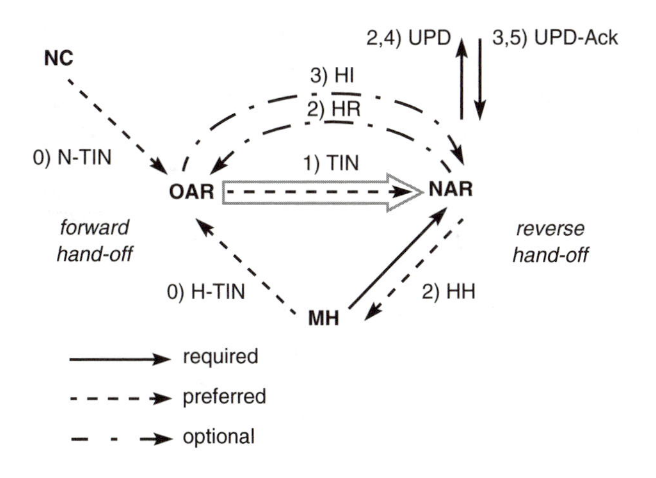

Fig 12.4 Basic EMA hand-off messaging.

by-hop towards the OAR, and may be re-sent multiple times until a UDU acknowledgement (UDUA) is received at the NAR or the OAR is determined to be unreachable. The NAR can also now send a hand-off acknowledgement (HAck) to the mobile so that it can start sending data over the new link. This message exchange remains the same for both BBM and MBB hand-off, whether or not the hand-off can be predicted.

12.3.2.3 'Anticipated' Break-Before-Make Processing

The message exchange used in the preceding example is now illustrated within a 'GSM-like' scenario of BBM hand-off (see Fig 12.5).

Only a portion of the message exchange and the three paths that packets can take to the destination mobile from the source mobile during hand-off can be shown due to space constraints. The H-TIN packet initiates tunnel creation and transmission of the TIN packet. The make event is seen in Fig 12.5, which initiates the UDU (the HR and HI phases are skipped here) to redirect routing via injection of host-specific routing state (shown next to the OAR prefix DAG state). Meanwhile, packets have been tunnelled towards the NAR since the break event. Packet flow is redirected in Fig 12.5 after the UDU hits the crossover router, and the tunnel comes down when the UDU hits the OAR. The UDUA phase is also omitted due to space limitations.

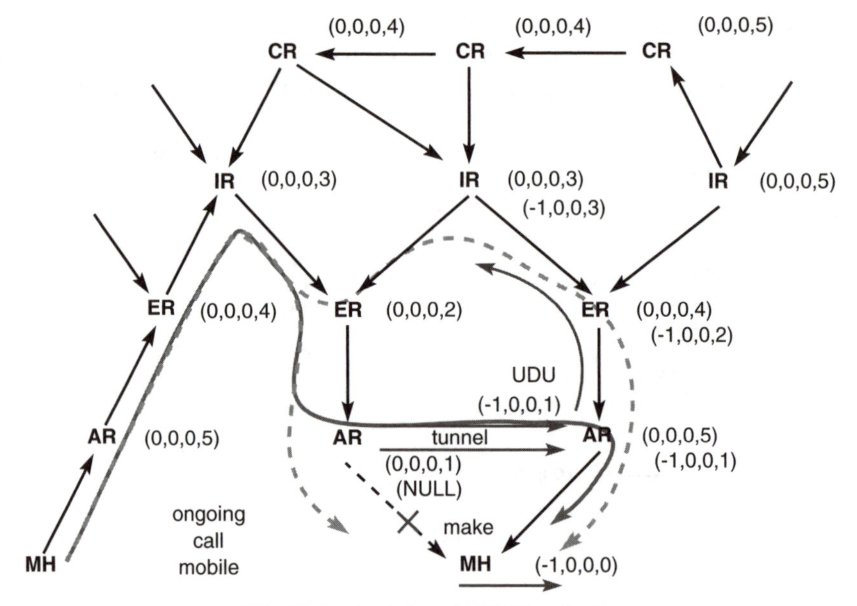

Fig 12.5 Anticipated BBM hand-off.

As the mobile migrates, a similar sequence will be repeated at each hand-off. Hopefully the reader will be able to construct similar message diagrams for the other cases involving unanticipated BBM, anticipated MBB and unanticipated MBB from what is presented here. These hand-off forms may occur as the mobile moves between different types of layer-2 technologies (e.g. GSM to Bluetooth hand-off) generating different hand-off event sequences. It should be remembered that it will be necessary to reallocate the IP address back to the AAR after a sequence of hand-offs. This is accomplished via a sequence of messages very similar to the hand-off processing (only in reverse) where the IP address is handed back to the AAR and all host-specific state is erased from the network which also cannot be shown due to space limitations.

12.4 EMA and Mobile IP Convergence

Mobile IP (MIP) (versions 4 and 6) provides for the potential movement of a home IP address throughout the Internet, from a home domain subnet, throughout foreign domain subnets equipped with MIP. It does so by providing a mobile node (MH) with a local and potentially short-term IP care of address (CoA) in a foreign domain, towards which packets can be indirectly tunnelled. This CoA is reported back to a home agent (HA), who then tunnels packets, sent by any correspondent node (CN) to the home address, towards this CoA.

In addition, directly tunnelled communication, bypassing the HA, can be achieved through additional signalling between the MH and the CN. A foreign agent (FA) entity can also optionally exist to assist the MH in the foreign domain by terminating tunnels and acting as a proxy CoA, a CoA allocator, and a proxy signalling point for MIP signalling.

The MIP collocated care of address (CCoA) is only a valid session address on a specific foreign link, and the utility of this address for native service is consequently severely limited, and its use therefore actively discouraged in mobile IP. Some movement of this CCoA address is, however, supported in MIP through the use of temporary tunnelling between ARs, with the current CCoA at the old AR being treated like a home address, and a new CCoA from the new AR acting as the tunnel end-point.

To aid subsequent discussions, temporary tunnel forwarding between adjacent ARs is described as 'horizontal' MIP forwarding, whereas standard indirect tunnel forwarding via the MH's HA in its home domain, or direct from a CN using optimisation, is termed 'vertical' MIP forwarding. This section outlines the interworking architecture for EMA and mobile IPv6/v4 that provides mutual co-existence and benefits, both for the user and the operator. The specific details change slightly between MIPv6 and MIPv4, and depend on the type of CoA (collocated or not).

12.4.1 Converged EMA/Mobile IP Addressing

EMA access routers provide a temporary address to the MH in any domain in which it roams, with the address changing on a per-IP session basis. This temporary address is allocated by the AAR when the MH has data to send, or when the MH has been paged for incoming traffic. In IPv6 the MH can automatically and rapidly build a temporary address by simply appending its EUI to the advertised prefixes from its chosen AAR. This temporary address (and associated service restriction) is similar to the address that is today allocated to the majority of dial-up users. These users do not need permanent IP addresses because they only utilise applications which are initialised through a client server process (Web, mail, newsgroups, ICQ, etc) in which outgoing traffic informs the servers (and any onward users) of the user's IP address. This temporary address is not, however, sufficient for users with servers who must be able to advertise the server address (e.g. DNS), for corporate roamers who require a valid address from their home network, and for users of peer-to-peer applications who wish to receive incoming calls.

Therefore, in addition to the temporary session address, the MH may also have a persistent IP address allocated via DHCP (or whatever) which is used at the IP layer as a globally unique, stable and hence advertised identifier. This IP address is associated with, and allocated by, the home ISP or corporate, and is associated with the home link of the MH. This address may be installed into DNS and SIP systems for example, and cached by knowledgeable users. When compared to cellular systems, the persistent address is in concept similar to the IMSI, while the temporary CCoA address is in concept similar to the T-IMSI.

The converged mobile IP/EMA addressing architecture requires that the mobile IP home address is used as the global, persistent address to ensure that the MH may be contacted when roaming using normal mobile IP mechanisms. The mobile IP CCoA is used as the local temporary EMA source address and enables this address to be used for MIP tunnelling as well as for native IP service. The benefits of this co-existence may be summarised as follows:

- mobile IP is used to provide roaming of a global address while EMA provides routable local addresses;
- the local address can be used by the mobile node to source traffic via any AR within the EMA domain due to the EMA host redirect routes and IP hand-off;
- the local address can also act as a sink of traffic from anyone in the Internet who learns the address (e.g. MH-initiated), limited then only by the scope of the address (global, site/link local, private, etc);
- the MH persistent home address can be used to sink traffic from any Internet host either via home agent or correspondent node tunnelling to the CoA;

- the MH persistent home address can be used to source traffic directly to correspondent nodes or via reverse tunnelling to the home agent and domain.

Further details of the benefits of this co-existence are described in O'Neill et al [10].

12.4.2 EMA and MIP with Collocated Care of Addresses

When the MH first starts up, the MH gets its first CCoA in this new domain and needs to undertake normal MIP signalling to register this CCoA with its (possibly remote) HA to facilitate indirect vertical tunnelling. In addition, the MH may wish to also communicate its new CCoA to any CNs so that direct vertical tunnelling (HA bypass) can be achieved. Forget, for the moment, what happened in any previous foreign domain, which may or may not have EMA:MER capabilities, and then assume that the MH is happily communicating via both direct and indirect vertical tunnelling when it detects a change of AR (link) using well-documented MIP techniques, possibly with help from the link layer.

An MER protocol would enable the CCoA to move with the MH as it changes ARs, by rapidly modifying the local routing tables so that traffic addressed to the CCoA now arrives natively via the new CoR/FA rather than via the old CoR/FA.

This is shown in Fig 12.6 and means that the CCoA and all MIP registrations are still valid, and hence there is no need either to update the vertical forwarding state in HA and CNs, or to undertake horizontal tunnelling between CoR/FAs. In an EMA:MER-equipped domain therefore, the MH only needs to acquire a single CCoA in that domain per IP session, and only needs to register that address once to the remote HA and CNs, as the MH moves in the domain. The result is faster hand-off, reduced signalling load, and in many cases no tunnelling of IP packets between adjacent ARs (CoR/FAs). This constitutes a significant performance improvement to MIP hosts if the EMA:MER domain is both practical, from a routing perspective, and sufficiently large to justify the deployment of EMA:MER. Note that, because of the clean separation between EMA:MER horizontal operations from the MIP vertical operations, and the focus of EMA:MER on optimising the horizontal plane (and as a result avoiding the side-effects on the vertical plane), each domain can independently decide on MER deployment without affecting vertical MIP operation as a user roams through foreign domains. From a MIP signalling perspective, the MH would still go through the motions of requesting horizontal MIP tunnelling. In a non-EMA domain they would be used to install the horizontal MIP forwarding tunnel as per normal MIP between the old CoR/FA and the new CoR/FA. However, in an EMA:MER domain, these messages would, instead of, or in parallel with, installing a temporary horizontal tunnel, be used to install the new host redirect route from the old CoR/FA to the new CoR/FA.

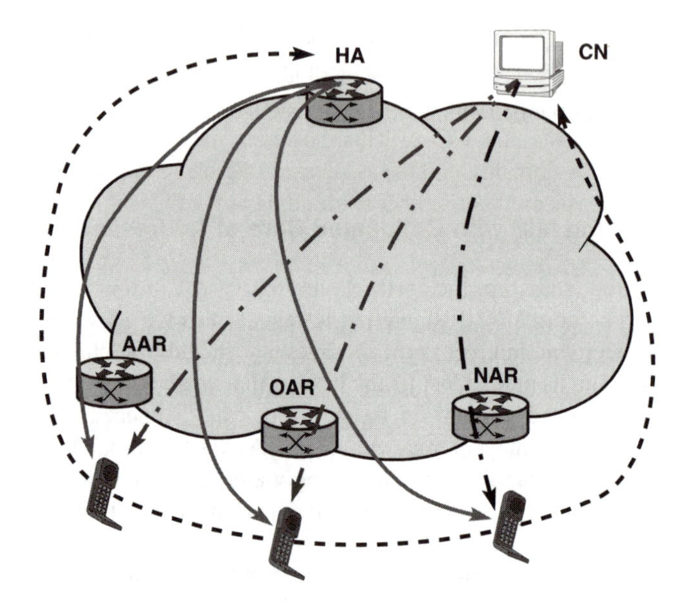

Fig 12.6 MIP updates in an EMA domain.

12.4.3 Inter-Domain Considerations

MIP is required in EMA between domains to support the movement of an MH IP address out of the local/home domain (no inter-domain MER), and is also required to support MIP hand-off in domains that are not equipped with an MER protocol. To facilitate the co-existence of MIP and EMA options, the MIP signalling capabilities with appropriate additions are obvious candidates for providing the signalling plane for EMA hand-off.

12.4.3.1 EMA MH Moves from EMA to Non-EMA Domain

When a mobile appears at an AR in a new domain, the present CCoA used by the MH from the old domain is no longer valid. The MH therefore needs to acquire a new CCoA in the new domain and find a way to continue to be able to use the previous EMA CCoA as a session address without leaving host routing state in the previous domain.

Present communications are using the old CCoA as a source address which are being routed by an MER host route towards the OAR in the previous domain. There are two hand-off stages to complete as shown in Fig 12.7.

- Stage one

  In the first stage the MH can send a binding update to the previous CoR/FA, as in standard MIP, so that it can horizontally tunnel packets addressed to the old CCoA forward to the new CCoA. This, however, means that the old CCoA (in the previous EMA domain), and the associated routing state from the AAR to the OAR, cannot be released until either the binding at the OAR times out and/or the route timer and address timer expire.

- Stage two

  In the second stage therefore, the MH now views the CCoA session address from the previous EMA domain as a secondary home address whose home agent is the AAR in the previous domain. This is called a collocated roaming CoA (CRCoA). The MH therefore sends another binding update (BU) to that AAR so that the AAR can AAA the request to roam with one of its CCoAs. If permitted, the AAR then provides HA and vertical tunnelling services to the MH for that CRCoA, and the MH can use BUs to CNs, using that CRCoA as a session address to bypass the AAR HA.

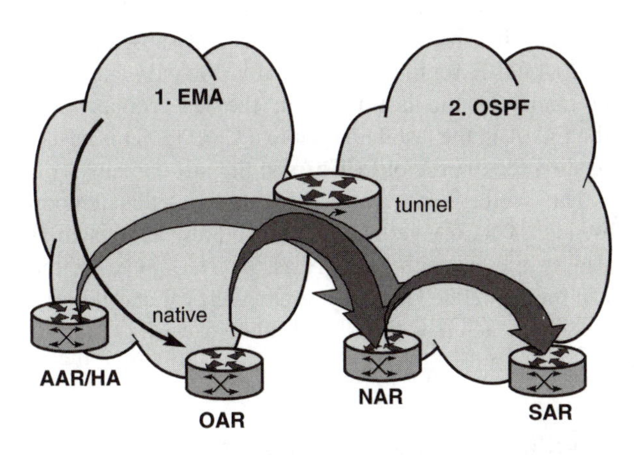

Fig 12.7 Inter-domain hand-offs using MIP tunnels.

The AAR will not need to act as the HA for the CRCoA and tunnel packets until the host routing state, directing packets addressed to the CRCoA towards the OAR, is eliminated on expiry of the temporary horizontal inter-domain tunnel. The AAR should not therefore return the BUAck immediately, but should delay completing this vertical tunnel set-up until the host redirect state has been removed. If the CRCoA request to the owning AAR is refused, use of the EMA CCoA is no longer possible and all sessions using this address will fail when the horizontal tunnel lease-time expires.

The destination of the MIP temporary horizontal inter-domain tunnel, and the vertical inter-domain tunnel for the new home address, is the NAR in the new domain. In parallel with this, normal MIP registration of the new CCoA for the original home address should also be sent to the HA of the MH to update vertical tunnelling based on the home address. This process will then be continued on subsequent hops as per normal MIP, with each new local CCoA allocation requiring both horizontal and vertical tunnelling updates for both the original home address and any acquired CRCoAs. An aggregated horizontal BU for the previous CoA and active CRCoAs, to the previous CoR/FA, would clearly be advantageous. Multiple CRCoAs are possible because a new CRCoA could be collected in each EMA domain encountered. This is not considered to be a significant scaling problem due to the rarity of inter-domain hand-off in the EMA model.

12.4.3.2 EMA MH Moves into an EMA Domain

If the MH supports EMA extensions, as soon as it steps into an EMA domain it will behave as if it is booting up in the EMA domain and will be required to undergo authentication. Then, the MH will be able to get a new CCoA from the AR to which it first attaches, which it can then continue to use on subsequent hops in the new domain due to EMA:MER routing. Additionally, the MH can choose to build an MIP horizontal tunnel to the last OAR in the last domain and then request conversion of the CCoA in the last domain into a CRCoA. This ensures that existing communications directed to previous CCoAs in the last domain, will not have to be dropped. Note that, while in the new EMA domain, this temporary horizontal tunnel, the subsequent CRCoA vertical tunnel, and any associated HA/CN vertical tunnels, will also be moved with the MH as they terminate on the MH's EMA-enabled CCoA in this domain. Therefore, a new vertical and horizontal tunnel will not have to be set up at each new CoR/FA, leading to vastly reduced signalling and faster hand-off.

12.5 Scalability Benefits of EMA:MER

Scalability is achieved due to a number of novel design decisions which should enable EMA to be supported across large ASs, equivalent to, but larger than, today's OSPF/IS-IS interior routed domains (many thousands of routers).

12.5.1 Temporary Session Addresses

The temporary session address means that an MH always starts any IP session being routable on a prefix route. A host route is only required for subsequent movement away from this first access router while in-session. When the session ends, the address and associated host state can be removed. When the mobile is not in-session, there is no implication on either the domain routing tables or the addressing resources.

12.5.2 Aggregate Prefix Routes

The use of prefix routes means that the network can grow to be a big domain with a large number of access routers, supporting both mobile and fixed access technologies. Mobility becomes a delta on top of the prefix fabric rather than having host routes replicate the job of the prefix route in a very inefficient way.

12.5.3 Localised Host Redirects

The presence of the prefix aggregate route, and the use of the temporary session address results in the host routing only being required for in-session mobility whose probability distribution has an impact upon the depth and breadth of host route injection, as shown in Fig 12.8.

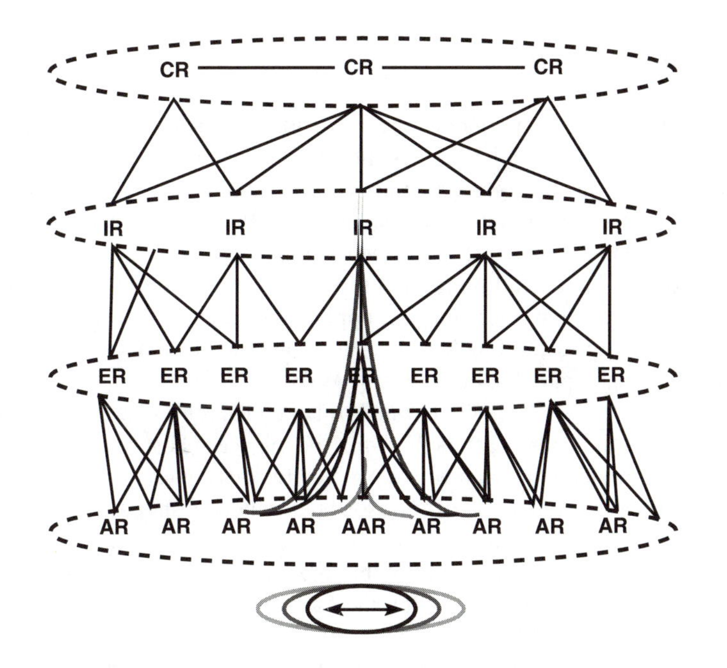

Fig 12.8 Movement probability away from AAR.

The directivity of the host redirect, between the NAR and the OAR, and the chaining of such redirects, results in significant localisation of mobility-induced routing messaging and host route state, instead of a domain-wide host route. The degree of localisation is dependent on the IP topology, associated routing metrics and even traffic-engineering-based TORA route selection.

The degree of router meshing and the number of levels in the router hierarchy affect the potential reach of redirect host routes. Higher degrees of meshing, lower in the network, result in a more direct horizontal path between adjacent access routers for the hand-off message and the resulting redirect host route. This causes incoming packets from the core, and from most other access routers, to take longer to reach the host redirect state. However, it will be seen sooner by packets coming from any intermediate AR (and associated connected sources) between the AAR and the present AR of the mobile due to the installed 'shallow' host route.

A more tree-like structure results in a triangular path between adjacent nodes going higher into the core. This will mean that incoming packets from the core and from most other access routers will see the redirect later while the intermediate AR packets will see the redirect earlier.

In addition, the use of preferential route metrics could be used to 'steer' the host redirect routes either higher or lower in the infrastructure as required, giving increased control above and beyond that 'implied' by the topology.

12.5.4 In-Session Dynamics

It is clear that an 'always on' mobile can be roaming anywhere over long periods of time. Any mobile routing solution that allocates persistent IP addresses is forced to track that address in the routing/paging system, leading to scaling problems. To overcome this, the paging system is separated from the mobile's care-of-address. In addition, the mobile is only tracked in the routing system during active IP sessions, which is likely to be a small proportion of the 'always on' mobile activation time. Further, while in-session, most Internet applications require that the user becomes relatively stationary in the topology because you cannot drive, walk or generally concentrate while, in parallel, browsing the Web.

There are certainly situations in which this application assumption is incorrect but in those cases it is other factors that limit the impact of host routes. For example, in the case of IP telephone calls where people are occasionally driving or walking, the call durations are short and the associated hand-off statistics are both well-known (GSM) and of little concern. In the case of planes and trains, such 'rare' host routes can be more simply supported within the localised cellular infrastructure and access routers along major routes through suitable dimensioning and associated tariffing. However, it is likely that, over time, IP infrastructure will be installed in such moving platforms, resulting in a single aggregated host prefix for all those collocated nodes moving in sympathy.

Further, the use of the mobile IP 'permanent home' address and bindings ensures that the user is always reachable if the host route needs to be removed (either route hold timer expiry or CCoA reset) [11]. Finally, the profile-specific IP session and route hold timers ensure that tariffing can be used to control the impact of in-session mobility and usage patterns on the overall routing system.

12.5.5 TORA:MER Scalability

In the most recent Internet Architecture Board (IAB) report on network layer issues [12], the IAB concluded that the scalability bottle-neck of presently deployed routing technology stems not from storage considerations but rather from long convergence times. These convergence delays are due to the time required to distribute stable routing information updates (communication complexity) and the time required to re-compute routing tables (computational complexity).

Operating in a suitable topology, TORA can have relatively low communication and computational complexity due to the nature of its distributed computation that forgoes shortest path computation.

12.6 Conclusions

This chapter has presented an approach for mobile enhanced domain-based routing that treats fixed and mobile-terminating traffic in an integrated fashion. It has the obvious advantage that many IP features designed for the fixed network will also work for mobile hosts. For example, the approach is amenable to traffic engineering above and cleanly separated from the routing function. By avoiding the use of long-term tunnelling, separate flows terminating at a mobile are visible and may be handled separately according to their traffic classes as part of a diffserv-based quality-of-service architecture. The approach also has the potential to replace or obviate the need for many layer-2 mobility signalling technologies, as well as to replace some existing IP routing protocols in fixed domains.

The work is now maturing with significant contributions being generated into the IETF, but simulation studies are now required for 3G topologies and traffic patterns. This concept combines three previously separate technology spaces — fixed routing, cellular mobility and MANET routing — and identifies a very promising direction for this fusion.

The approach hinges on the use of flat, non-shortest path routing in a large-scale, hierarchical mesh topology. Control of topology capitalises on the strengths of the routing algorithm while diminishing its weaknesses. For example, a sufficient degree of meshing within the topology permits exploitation of the algorithm's localisation properties, thereby avoiding far-reaching control message propagation. Freed from this convergence bottle-neck, the topology can be made larger, potentially much larger than is possible with traditional shortest-path routing technology. It is apparent then that obtaining the best performance requires a trade-off — a proper balance of topology design and control during hand-off to fully exploit the algorithm's localisation properties while preserving routing efficiency.

Future work will focus on several areas. The current hand-off model is highly localised, involving only the MH, the OAR and NAR and the set of routers on the preferred paths between OAR and NAR — in this sense similar to the hand-off model of HAWAII [13]. Localisation is desirable for scalability in that control messaging is minimised. However, in mesh topologies there are costs associated with this degree of localisation. As the mobile continues moving, it will leave a trail of host-specific state in the network. Also, in certain topologies, the routing may still be less optimal than desired due to node mobility. Additional hand-off mechanisms are under consideration as performance enhancements to address these issues, and significant simulation study is needed. Development will also cover the mechanisms needed to handle dynamic topologies for the edge mobility enhancements to TORA, and there are plans to put forth a companion paging architecture.

References

1 Valko, A.: '*Cellular IP: A new approach to Internet host mobility*', ACM Computer Communication Review, **29**(1) (January 1999).

2 Campbell, A., Gomez, J., Wan, C. Y., Turanyi, Z. and Valko, A.: '*Cellular IP*', Internet-Draft (work in progress), draft-valko-cellularip-01.txt (October 1999).

3 O'Neill, A. and Corson, M. S.: '*Edge mobility architecture*', Internet-Draft, draft-ietf-oneill-ema-02.txt (July 2000).

4 O'Neill, A., Corson, S. and Tsirtsis, G.: '*EMA enhanced mobile, IPv6/IPv4*', Internet-Draft, draft-oneill-craps-handoff-00.txt (August 2000).

5 O'Neill, A., Corson, S. and Tsirtsis, G.: '*State transfer between access routes during handoff*', Internet-Draft, draft-oneill-handoff-state-00.txt (August 2000).

6 Park, V. and Corson, M. S.: '*The temporally-ordered routing algorithm*', Proc IEEE INFOCOM '97, Kobe, Japan (April 1997).

7 Park, V. and Corson, M. S.: '*Temporally-ordered routing algorithm (TORA) Version 1 Functional Specification*', Internet-Draft (work in progress), draft-ietf-manet-tora-spec-02.txt (October 1999).

8 Park, V.: '*A highly adaptive distributed routing algorithm for mobile wireless networks*', Masters Thesis, University of Maryland, College Park, Maryland (1997).

9 Mills, D.: '*Network time protocol, specification, implementation and analysis*', Internet RFC-1119 (September 1989), and RFC-1305 (March 1992).

10 O'Neill, A., Corson, S. and Tsirtsis, G.: '*EMA enhanced mobile, IPv6/IPv4*', Internet-Draft, draft-oneill-ema-mip-01.txt (July 2000).

11 Perkins, C. E.: '*IP mobility support*', Internet RFC 2002 (October 1996).

12 Kaat, M.: '*Overview of 1999 IAB network layer workshop*', Internet-Draft, draft-ietf-iab-ntwlyrws-over-01.txt (October 1999).

13 Ramjee, R., La Porta, T., Thuel, S., Varadhan, K. and Salgarelli, L.: '*IP micro-mobility support using HAWAII*', Internet-Draft (work in progress), draft-ietf-mobileip-hawaii-00.txt (June 1999).

13

3G TRIALS AND DEVELOPMENTS

C J Fenton, J G O Moss, D W Lock, R Bloomfield, J F Fisher, D T Pratt, A Brookland and J Gil

13.1 Introduction

For a long time the wireless data wave has been muted as the biggest change in mobile service offerings that will open up new opportunities. With the launch of GSM data services the data wave was said to be beginning to build up ready to wash into the market-place — but it did not happen. Many factors were missing, the most important of these being the basic data rate was too low and there was a lack of consumer terminals available to the user.

At GSM's conception, discussions of two bands of services (classed as 'slow' — below 4.8 kbit/s, and 'fast' above 4.8 kbit/s) were promoted widely. By the time these came to the market, all were clearly 'slow' and, with modem technology in the home offering 28.8 kbit/s, this clearly made the GSM offering unattractive. 3rd generation (3G) had to overcome these basic issues especially as the World Wide Web was pervading our everyday lives and corporate intranets had become the key mechanism for information storage and dissemination.

Therefore the key questions on everybody's lips were: 'What are we really going to do with this 3G capability?' and 'How much will it cost?' With the hindsight of the GSM experience a trials activity was planned to examine these two basic questions. The plan concentrated on validating the radio system that promised radically different data rates, and on the service offerings that were made possible through the use of IP. More recently the project work has been branded the 'RoadWarrior' and expanded to explore corporate access and personal mobility through automation of access.

What was needed was an approach that would enable an innovative hot bed of creativity. The team discussed the approach and drew out the essential qualities that were needed:

- to demonstrate multimedia applications hands on;
- to build on real devices to assess potential and shortfall of device capability;
- to integrate vendor capabilities so as to develop alpha products;
- to workshop alpha products with the technical and marketing team;
- to characterise vendors' radio systems so as to assess for system planning;
- to review design specifications for validation of standards values.

Figure 13.1 shows the various streams of work and the relationships between them to take the trials work forward.

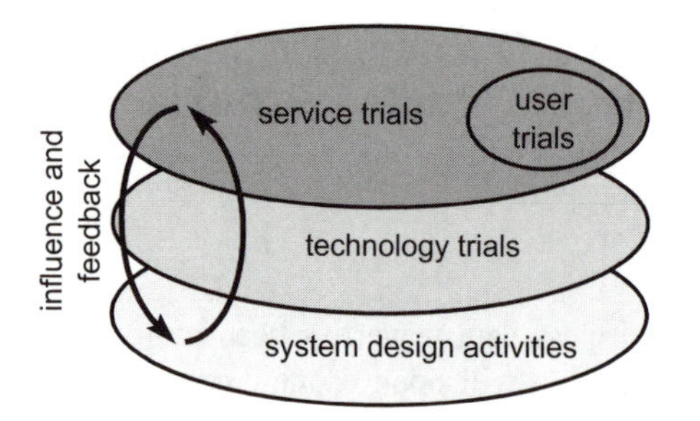

Fig 13.1 Trial work streams and relationships.

The specific objectives of the trials stage are best summarised as:

- to demonstrate applications and act as a catalyst for new revenue-generating product ideas;
- to create a permanent environment to link all the technology elements (work with vendors);
- to validate and evolve system design requirement parameters leading to product/ feature specification;
- to begin to live the 'IP world of 2004'.

13.1.1 Service Trials

The service trials formed the realisation of the vision — to present the vision in a tangible, stimulating way that could represent both a marketable product concept, and demonstrate the capability that underpinned the product. In this case, the capability referred to is the enabling application (or sub-layers) that brings the service to life over the access system below. This activity also enabled vendors with new applications to undertake testing in the mobile environment. The system was an all-IP local area network (LAN) with multiple sub-domains. It was put in place and could operate without other systems or interfaces so that progress could be made as fast as was needed.

This was particularly useful as the technology trials section was very dependent upon commercial relationships.

13.1.2 Technology Trials

The technology trials formed the realisation of a number of access technologies. This activity provided real access systems for trial and demonstration of the applications/services over real systems. A number of vendors were approached and, although agreement was reached in a number of areas, the delivery was difficult to manage with uncertainty surrounding the delivered capabilities.

13.1.3 System Design Activities

With the benefit of hindsight from the GSM standardisation and from the network commissioning work that some of the team had been involved in with Cellnet, having the ability to set parameters or to validate parameters referenced in standards activities is key to a successful standards and equipment implementation. The work here was to go hand-in-hand with the system design teams, where appropriate, to ensure that parameters (e.g. QoS) levels were correct.

13.2 Radio System Trials

The target 3G network consists of a number of elements, much like the GSM system, but with some differences and with new device names to differentiate it from GSM (Fig 13.2).

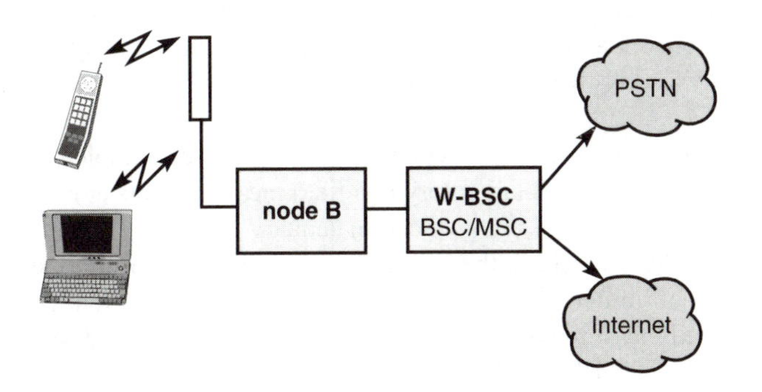

Fig 13.2 Generic UMTS radio system.

Systems offered for test were early developments and consisted of simulated network components but with radio systems that offered the characteristics of the launch UMTS system.

13.2.1 Chronology of UMTS Trials Equipment

FDD UMTS prototype trial equipment based upon the ARIB standard was supplied by four companies as shown in Table 13.1.

Table 13.1 Prototype equipment suppliers and testing chronology.

<table>
<tr><th>Supplier</th><th>7/98-12/98</th><th>1/99-6/99</th><th>7/99-12/99</th><th>1/00-6/00</th><th>7/00-12/00</th></tr>
<tr><td>Ericsson</td><td></td><td colspan="3">Lab and drive testing</td><td></td></tr>
<tr><td>NEC</td><td>Phase 1</td><td>Install</td><td colspan="2">Phase 2</td><td>Phase 3</td></tr>
<tr><td>Nokia</td><td></td><td></td><td>L&D</td><td></td><td></td></tr>
<tr><td>Nortel</td><td></td><td>Install</td><td>Drive testing</td><td></td><td></td></tr>
</table>

All the testing carried out on the Ericsson system was undertaken on their test network based in Guildford. The scope of the testing was determined by the functionality available as the system underwent upgrades. A 'node B cabin' was installed on the Adastral Park site in January 2000. This enabled the demonstration to BT internal customers of Ericsson-specific applications.

The NEC trials have been split into three phases. Phase 1 which commenced in October 1998 consisted of two identical transceiver units, configured as base-station and mobile station. Basic air interface measurements were undertaken on this equipment. Phase 2 (June 1999) consisted of two base-stations working into an MCC simulator. This combined the basic functionality of a base-station controller (BSC) and mobile switching centre (MSC). Laboratory and field-based measurements were undertaken between June 1999 and June 2000. Phase three of the trial has primarily consisted of subjective application testing. The basic transmission network associated with Phase 2 has been enhanced by the provision of IP connectivity to BT's RoadWarrior platform and to the Internet via the appropriate 'fire wall'.

The Nokia trial commenced in September 1999 and was completed in December 1999. This consisted of both laboratory and field measurements at Nokia's test sites in Helsinki. The Nortel trial commenced in January 1999 with the installation of a trial network at Adastral Park; the system was not, however, fully operational until June 1999. The limited functionality from the data collection tool prevented laboratory-based measurements. A number of parameters were measured during the drive testing which finished in December 1999.

13.2.2 WCDMA Laboratory Radio Trials

The WCDMA laboratory radio trials commenced in October 1998 with the delivery of two basic WCDMA transceivers from NEC. This provided the opportunity to test and gain experience of the new technology and derive a set of fundamental radio tests with which to determine the spectral efficiency of the systems. Under investigation were some fundamental planning parameters, particularly a receiver noise figure and the required signal-to-noise ratio (SNR) to support a required bit error rate.

WCDMA can support a variety of data rates from speech at 12.2 kbit/s and data service bearers up to, and beyond, 384 kbit/s. The signals from individual users are multiplexed using spreading sequences and matched filter receivers. As a result, accurate up-link power control is necessary, based on the required SNR. Accurate performance of this operation is critical to the improved spectral efficiency of WCDMA compared to GSM.

The required SNR is dependent on the user data rate selected and the speed of the mobile. The building topography around the mobile also affects the necessary SNR due to the different multipaths that exist between the base-station and mobile. Laboratory emulation of such radio channels was performed using radio channel simulators, which accurately create the required multipaths and path attenuations. The tests used representative and measured micro-cell and macro-cell radio channels as well as single-path static channels, which represent a best-case condition. As a result, indications of the required SNR for each user data rate for a range of possible environments and mobile speeds were produced.

Further tests confirmed the accuracy of the up-link power control to maintain this SNR despite the time-varying channel. The overall combination of these two parameters produced an indication of the likely Erlang capacity of each cell for a combination of services.

Further tests were performed to determine the interaction between adjacent WCDMA carriers. WCDMA carriers are spaced at intervals of around 5 MHz, and each occupies 4.7 MHz of bandwidth. As a result, it is important that minimal signal power 'spills' into adjacent bands due to nonlinearities within the transmit amplifiers. Similarly, each receiver must be able to accurately filter out all adjacent band interference to extract the wanted signal. These effects introduce additional interference into the network, which can cause calls to be blocked, if sufficiently severe. The above effects were quantified for the systems within the laboratory to provide indications as to the likely adjacent channel interactions.

The evaluation showed that the systems performed close to the predicted values, thus validating the standard. The figures were also used to calibrate modelling tools for network layout and cost estimations. These figures were used to assist the 3G licence bid team in their spectrum licence valuation.

13.2.3 Wide Area Installations and Trials

Ericsson, NEC and Nortel all supplied base-station equipment for evaluation at Adastral Park.

The Ericsson installation consisted of the base-station equipment housed in a self-contained cabin. The cabin was installed on a prepared concrete base. An antenna system consisting of a three-sector dual-polar array was installed on the existing tower. A point-to-point 18-GHz microwave radio link between the test range and radio tower provided the local interconnection to an existing line system for a 2-Mbit/s (E1) circuit to the controller.

Traditional three-sector, space-diversity, antenna systems were installed on the radio tower at Adastral Park and also on Foxhall telephone exchange located on the outskirts of Ipswich (see Fig 13.3).

Fig 13.3 Antenna installation at Foxhall (left) and Adastral Park (right).

The installation on the tower was used for both NEC and Nortel at different times during the trial periods. The Nortel BTS and MCC-Sim were located in a room close to the main demonstration to enable radio coverage to be obtained (Fig 13.4).

Due to the limited accommodation available at Adastral Park, the second NEC BTS and MCC-Sim were located at the telephone exchange. An 18-GHz system was also used for the interconnecting links.

These prototype systems have provided some interesting 'opportunities'. For a fully equipped three-sector base-station, the power consumption is approximately 20 amps (at 48 V) per sector. The requirement for the high current is due to the need for the transmit power amp to be linear across the operating frequency band. The PSUs were obsolescent units kindly supplied from BTCellnet redundant stock. The transmissions between the BTS and MCC-Sim of both the NEC and Nortel systems are carried on a J1 bearer (1.5 Mbit/s). Proprietary interface boxes were used which provided a J1-E1 interface. When the trials commenced in late 1998, there were no

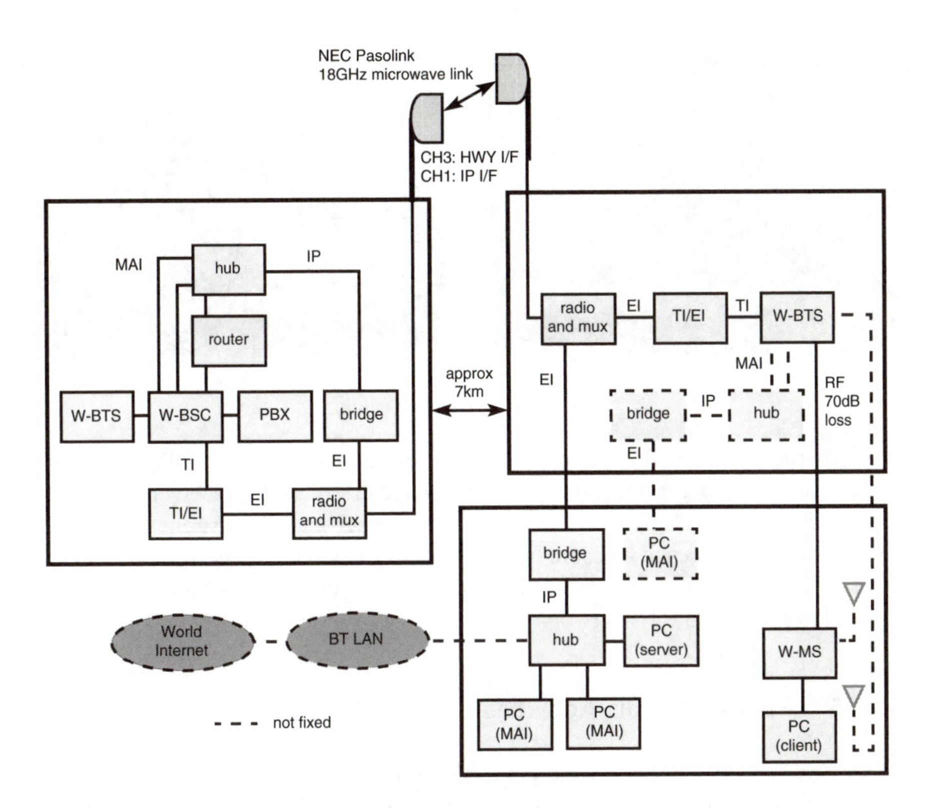

Fig 13.4 Connectivity across various sites and buildings.

commercially available antennas designed to cover the proposed FDD part of the UMTS band. Therefore a small quantity of prototype single-band single-polarity equipment was used. Suitable brackets had to be designed for fixing to the steelwork structures.

13.2.4 Drive Tests

Drive tests are necessary to prove the equipment in a truly mobile environment and to collect data that can be used to calibrate the radio planning tools. To enable the tests to be carried out it is necessary to set up a test network and to have a test mobile available to make measurements of the air interface. The test mobile is connected to a PC, which is used to collect the required data, and both are mounted in a drive test vehicle. As the test mobile is power hungry (28 amps at 12 V), a specialist test vehicle based on a Land Rover Discovery (Fig 13.5), with adequate power supplies and internally fitted differential GPS with heading sensors, was

Fig 13.5 Drive test vehicle.

used. As the vehicle is driven around the chosen test route, data is collected by the PC and can be plotted against the position collected from a GPS receiver.

The data collected gives information on the received signal strength from the base-station, interference measurements, and information on the performance of the system. This is then used to model the propagation prediction in the tool, which will be used to plan the UMTS network.

13.2.5 Future of Trials Activities

Much has been mentioned in the press regarding the costs involved in obtaining UMTS licences. The spectrum package awarded to BTCellnet in the UK included a 5 MHz block allocation for time division duplex (TDD) use. TDD uses the same frequency for both transmit and receive with the ratios of up-link to down-link time-slots being asymmetric, thus allowing higher data rates in the down-link. This technology is suited to in-building and 'hot-spot' pico-cell use. Work commenced in the autumn of 2000 to obtain basic coverage and capacity figures. This initial work is being undertaken at Adastral Park using a Siemens prototype system.

13.3 Application and Service Trials

In order to gain greater understanding of 3G systems and potential services, a hands-on development/integration system was constructed. The aim of the activity was to create an advanced platform providing a realistic view of 3G network capabilities, to demonstrate 3G service scenarios, and to develop the functional network elements to support these scenarios. By implementing an example of a future 3G network, an increased understanding of the issues and opportunities is being gained.

The platform has also been used to demonstrate BT's vision of future services and applications to a wide range of audiences.

13.3.1 Platform Development Focus

3G networks are about much more than just high-speed wide-area cellular access. The BT Group vision for 3G defines a new customer experience: 'A personal-profile-based service encompassing information, communications and entertainment which are accessible in a consistent manner at any time, anywhere.'

The distinction between fixed and mobile will converge as subscribers take advantage of advances in both personal and terminal mobility. Therefore the development of the network elements required to deliver 3G-type services has concentrated on constructing a platform that is access-method agnostic.

3G systems will bring true user mobility. Subscribers will be able to roam seamlessly between different networks using different access methods. The best way to realise this vision is to base 3G systems upon IP service delivery. The primary reason is that this will provide the ability to run IP services over a range of underlying access and transport mechanisms.

13.3.2 Platform Functionality

In order to build a realistic model of a 3G system, the functional elements required to provide 3G-type services need to be identified (Fig 13.6).

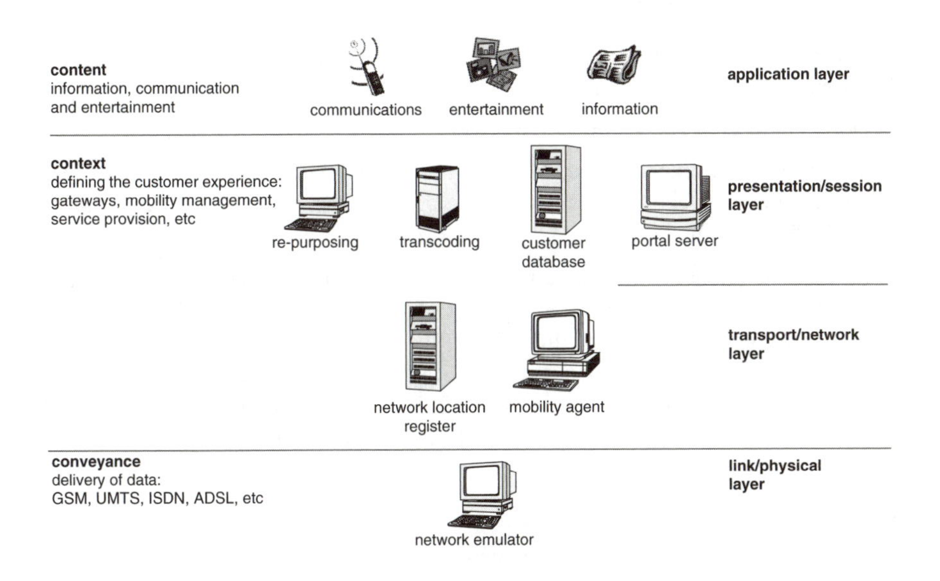

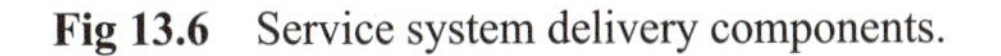

Fig 13.6 Service system delivery components.

- Content

  3G services will encompass not just voice and data but all subscriber information, their communications and entertainment.

- Context

  These elements define the customer experience ensuring user content is relevant, timely and formatted appropriately for the user terminal and network capabilities. Network elements must carry out functions such as mobility management, subscriber profile storage and content formatting and repurposing. This area has been the main focus of the activity.

- Conveyance

  Although the development platform has been designed to be access-method agnostic, a wide range of networks and terminals will be available on 3G networks. This range of capabilities, such as network bandwidth and delay, or terminal screen size and multimedia support, will have a great effect on service provision. To assess the effects, both a range of user terminals and a network emulator will be used to simulate future networks that are not yet available (see section 13.3.5). Considering the factors described above the following main functional elements were chosen for development.

13.3.3 Platform Operation

With the functional elements described above implemented, the platform is able to demonstrate the following concepts.

- Domain mobility and seamless handover

  Subscribers will expect to be able to roam not just within the wide area cellular domain but between different networks such as the combined office intranet/ VoIP PBX and the home voice/data network.

  In order to demonstrate domain mobility, the platform has several networks which are defined with the development platform as one of the following types — home, mobile, office, public. Within the development platform, mobility is provided at the IP level using mobile IP (see section 13.3.5). This gives the conceptual architecture shown in Fig 13.7. By providing mobility at the IP level, seamless handover is possible between the different conceptual networks available on the platform.

- Location management

  In order to provide personalised content to users, services must take into account the user's location, network type and terminal type. In order to obtain this information a network location register is populated with the subscriber's network address identifier (NAI) and a mobile IP care-of address (CoA) retrieved

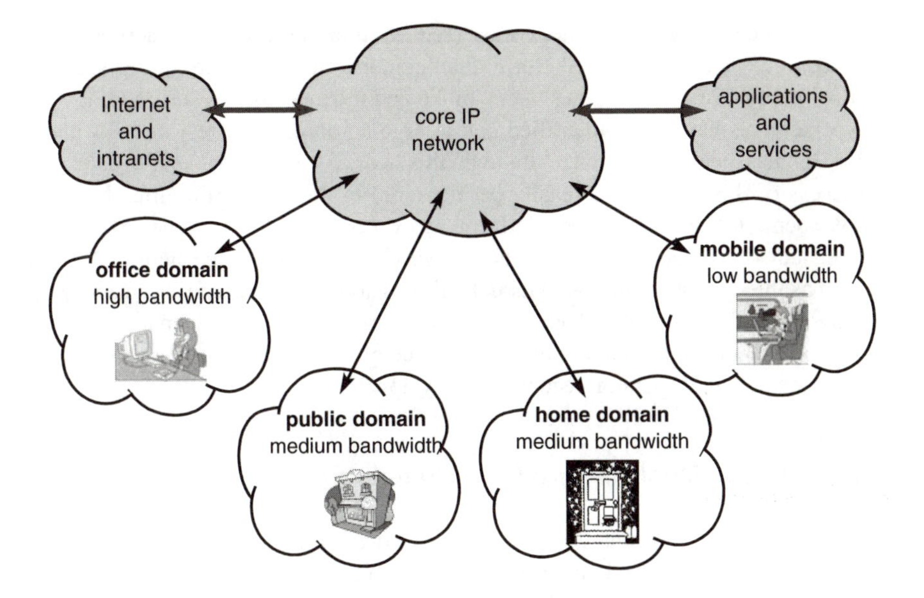

Fig. 13.7 Domain categorisation.

from the mobile IP home agent. The system correlates the subscriber's current mobile IP CoA with a specific network domain and characteristics for that network, e.g. bandwidth. This information can be used by services and applications to determine appropriate content and formatting.

- Personalised content provision

  As there is a vast amount of information freely available on the Internet, how can revenue be generated from content provision? In order to generate revenue from these types of services, content needs to be made valuable to users, ensuring it is relevant, timely and formatted appropriately for the user terminal and network capabilities.

  Subscriber profile data (such as to which services users subscribe, which content, e.g. news, they read, and interface settings) is stored in the system's central database. The platform uses the profile information along with the subscriber's location, and, using the network capabilities, provides profile- and location-dependent content via a Web portal. As users move around the network, the portal automatically provides updated content relevant to the user's current circumstances without the need for user interaction, such as reconfiguration, etc.

- Content repurposing — terminal and network

  Adapting content from outside an operator's network is more difficult but can still be achieved to a degree, such as removing or transcoding images to improve access speed over slow networks (see section 13.3.4) This activity has successfully developed a platform that demonstrates 3G-type services using 'real' network elements rather than 'smoke and mirrors'. Issues surrounding 3G-type services have been identified and a large amount of interest in BT's plans for 3G has been generated. Although this work is relatively small-scale, the success of this activity can easily be measured by looking at the time the team has spent demonstrating the platform at events around the globe. Many 3G workshops have been based around the service visions created here and often used as the opening stimulus to take the discussions forward. The service visions provide confidence that IP-based technologies can deliver more than a best-effort service — indeed the threat that other service providers can take the value away from the network operator becomes starkly apparent.

13.3.4 Use of Mobile IP and Active Filtering

As 3G cellular network technologies improve, the level of service which can be offered across the mobile domain increases with the bit rates available. However, the notion of seamless services across a variety of different network types is becoming increasingly important. The goal of an 'always-on' network link to a mobile device, via whatever network access route is currently available, can now be achieved. Devices accessing an IP backbone data network can expect a seamless service provision across multiple access routes, be they cellular, wireless or fixed.

The provision of content-push services and this desire to maintain a commonality of services across multiple access routes, bring forth a need to produce terminals with intelligent network support in order to fulfil the original ideals behind the UMTS vision.

The notion of an always-connected terminal, however, brings a series of additional problems. One of these is the notion of information delivery. While a terminal can retain an IP connection across multiple access networks, those networks are likely to vary considerably in the level of service which they can reasonably support. For example, while a fast wireless LAN connection such as HIPERLAN can comfortably support a high-definition streamed video link to a mobile terminal, a low-bit-rate GSM link would instantly become congested under the same data load. This leads to a need to ensure that content is adapted to best suit the network currently being utilised by a mobile user (Fig 13.8).

13.3.4.1 Objective

This issue leads to a desire to produce a terminal which can access multiple networks via an all-IP backbone, together with mobility support within that network

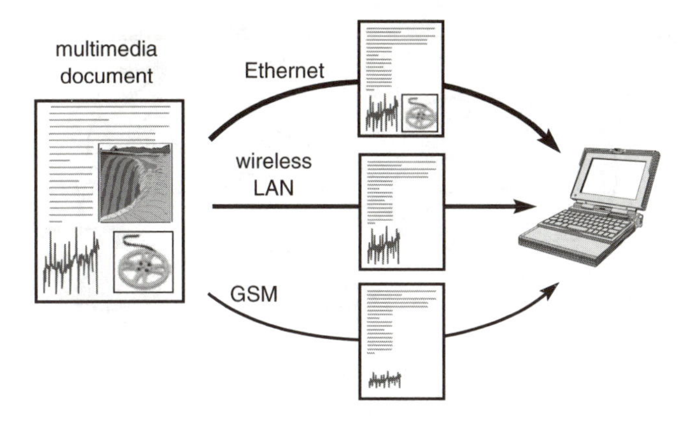

Fig 13.8 Content to match network.

to provide a seamless service provision with appropriate content adaptation in order to ensure that content delivered is suitable for the current network.

13.3.4.2 Solution

The solution to the above issues was developed, in some measure, as part of the M3A (mobile multimedia access) ESPRIT project within BT, and subsequently downstreamed to provide the network backbone for work within RoadWarrior.

Firstly, an internally developed mobile-IP-based solution to the problem of network mobility was adopted to solve the problem of seamless IP service provision described above. This allows mobile devices to maintain a fixed address within an IP core network, their home address, and acquire local temporary addresses on remote sub-nets as they roam around, either through the use of foreign agents, or by allocating collocated addresses in advance. Any traffic destined for that mobile terminal is then routed via their home address, where it is intercepted by a home agent. The traffic is then encapsulated within a new packet for delivery to their new temporary home and de-capsulated for local delivery either by a foreign agent, or by the mobile terminal itself (Fig 13.9).

This mobile-IP base also provides the RoadWarrior platform with information about a roaming user's current location and network access method. This allows Web-proxy-based transcoding services to provide content repurposing services as required. This might mean pointing requests from some users to pre-prepared low-bit-rate services, passing data through a transcoding engine, such as those provided by SpyGlass for WAP-based data for example, or caching files locally for later delivery if the currently available network cannot support their delivery.

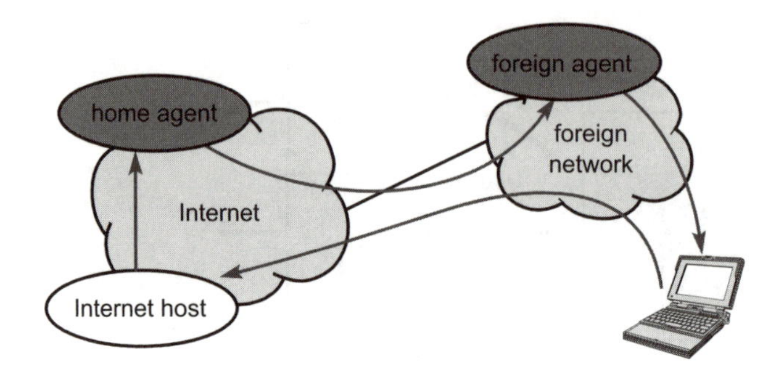

Fig 13.9 Seamless IP service provision.

13.3.5 Cellular System Channel Emulator

The idea of this system was to provide a test-bed scenario for application developers to try their applications before launching the software into the market. It was necessary to create a simple device which provided a first approach on how a given application would behave in the real network. After some months of work, the solution used was a PC-based router with a Unix-based operating system, at least two different network interfaces, and some software for simulating different networks.

The emulator (Fig 13.10) basically handles packets at the IP layer, using a program from the National Institute of Standards and Technology. Networks are characterised by some parameters, like one-way delay (either fixed or variable), bandwidth, bit-error rate, and packet duplication. IP packets are then handled according to these parameters.

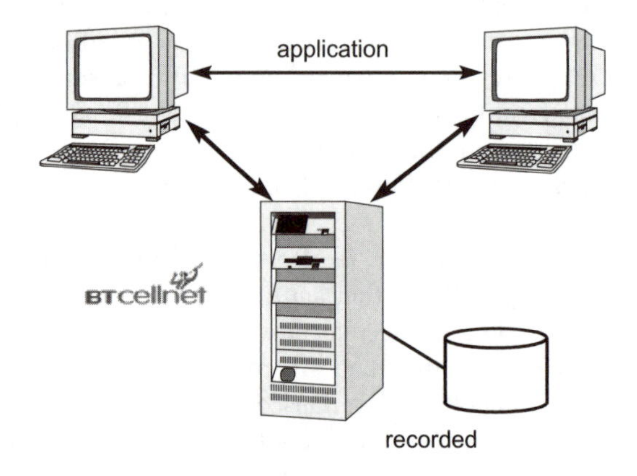

Fig 13.10 Initial emulator system.

It is also possible to record a ‘route’ where handovers and change in data rates can be replayed as though the user were driving along. The system allows both recording of routes and creation of worst-case changes to data bearers that then allow better application testing to be undertaken.

In order to test client/server applications, client and server are both connected on different network interfaces to the emulator. With this simple configuration, both wired and wireless IP-based networks can be simulated. It is also possible to simulate radio characteristics like changes of cell (handovers) and low-coverage situations, as well as varying the conditions of the network dynamically. In this way, application developers can have a first view on how the application will respond in different conditions, allowing modifications that otherwise would be expensive to do once their software is in the real world.

Users interact with the emulator via a graphical interface which enables them to add together as many networks in the data path as their target market sector requires.

Figure 13.11 shows two scenarios that represent the target application system. It also shows the graphical interface which allows ‘network’ values to be strung together to create an overall system experience for the application.

The system is being used by BTCellnet in Slough to emulate the new GPRS network, and at Adastral Park to test some video, audio and browsing applications, testing different video and audio codecs to see how robust they are under different emulated conditions.

It is being developed to create a better representation of the layers provided by GPRS so that application protocol interaction, where there may be conflict or cumulative delay caused by retransmission algorithms, can be assessed and ‘soak’ tested. Developments for terminal devices that have different physical connectivity (i.e. IrDa or Bluetooth) are being developed so that the application can be operated on the target device and repeatable system emulation undertaken.

13.4 Trials Outcome

So where has all this work got us? Overall the activities have matured our understanding and provided evidence that the combination of higher data rate with the very flexible capabilities of IP-based implementations will create an exciting new communications opportunity. The data wave can be seen to be taking off at last. All the people that have visited and talked through the demonstrations have all found the experience eye opening:

- it brings mobile multimedia to reality through tangible imaginative service offerings;
- it raises issues about how to deliver services to small devices where the screen is only 2 × 3 inches with relatively low resolution requiring none of the 2 Mbit/s data bandwidth;

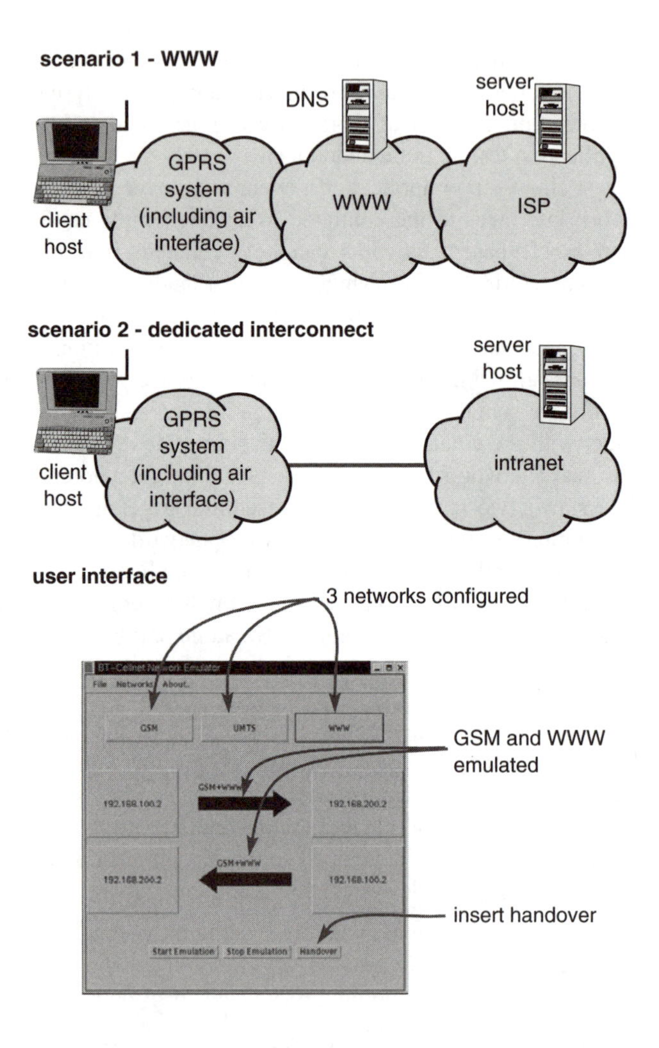

Fig 13.11 Emulator scenario and user interface.

- it brings out how IP can be used to deliver a usable service quality and how it still has features to be standardised or implemented for best service delivery;
- network-based features and functionality are identified through the 'real' implementation and options for system architecture discussed — the appreciation of the threat from other potential service providers to deliver via their components outside the operator's network are brought to light;

- the need for a different skill set and system appreciation for opportunity is given some bounds and plans to change the structure of organisations to meet the new market created;
- it raises the operations, administration and management differences and challenges (e.g. user traffic data collection).

13.5 Conclusions

There are still major areas that need to come together in order to make the real service happen. By far the most significant of these is the terminal device. Manufacturers have very difficult product line decisions in front of them as each sector requires a different terminal which has highly complex protocols and implementation issues within it — it is not cheap to make a small number of terminals and therefore mass-market production is the only way forward. So we may well see the terminal manufacturers create both highly successful winners as well as very disappointing public failures. The costs could be large enough to break the manufacturer and it may well be that only a few survive who then go on to create a family of devices with user interfaces which become as familiar as MS Windows on your PC.

Devices need to range from affordable communicators for mass markets like 'pre-pay', through to highly capable portable personal computers for those with specific needs. However, we will see for some years to come a continual revolution in terminal devices as large-scale integration of devices means fitting more functions into smaller spaces. In some of the demonstrations the line '.....data wave goes wireless.....' is used and the experiences that the user will have with the new system and devices are itemised. The trials work has highlighted the opportunity and the items that still require development in order to make this happen.

Currently the network capability and the 'all-integrating technology' IP are being put in place — there is therefore nothing to prevent this opportunity being seized and the world of which we have so easily talked being created.

14

PROFESSIONAL MOBILE RADIO — THE BT AIRWAVE SERVICE

P R Tattersall

14.1 Introduction

Professional mobile radio (PMR) has a much lower profile than public cellular, mainly because it was never aimed at use by the general public, but it was and is a professional tool, indispensable to a large number of 'group'-oriented users. These users are as diversified as police, fire, ambulance, industrial and other commercial users such as transport and taxi companies (Fig 14.1).

Fig 14.1 PMR standards and technologies are built for group communication. Public safety users have mission-critical workforce and incident management requirements.

The standards for TETRA have been developed at a time when the capability of GSM was being greatly enhanced through developments such as GPRS, CAMEL, EDGE, etc. The main reasons that public cellular technology was not suitable for mission-critical 'public safety' use are that only PMR offered the required functionality and performance, and that a very high level of service had to be guaranteed. The latter is a somewhat different proposition to current cellular service provision — even UMTS, the 3rd generation public cellular system, will not offer this level of service. Some key examples of current PMR TETRA functionalities are:

- fast call set-up (often using press-to-talk (PTT) operation);
- group and broadcast calls;
- despatch operations ('command and control');
- packet data;
- direct mode (users can communicate with each other directly in the absence of coverage from the radio infrastructure);
- security (over-the-air and end-to-end encryption, as well as mutual authentication of users and network);
- user-selectable priority levels with pre-emption for emergency calls.

The Airwave system is based on the ETSI TETRA standard for PMR. This is a scalable mobile technology allowing cost-effective implementation, from individual sites, through mid-scale systems (such as the London Underground and West Midlands Ambulance), to national systems (such as the Airwave public safety and Dolphin commercial services).

14.2 BT Airwave Service Overview

The 'blue light services', e.g. police, fire and ambulance, have particularly demanding functional and performance requirements, which can only be met by bespoke PMR network design.

An intensive 'project definition study' of requirements and specifications in 1998 led to the Airwave [1] service design being accepted last year by the Police Information Technology Organisation (PITO). The basis of this contract was the supply of a public safety service under PFI/PPP (Private Finance Initiative/Public Private Partnership) terms.

At the time of writing, the initial 'pilot' stage is well under way, and the network will be rolled out throughout England, Wales and Scotland by 2005. Service contracts will be agreed with each force for fifteen years, which means that the Airwave service will still be operational in 2020 (Fig 14.2). There will be many advances in technology and service requirements over this time, which is why a clear path to enhancing existing technology is needed, even though the initial service is only just being implemented.

The Airwave service provides secure and reliable voice and data communications, leading to better and faster provision of information to mobile

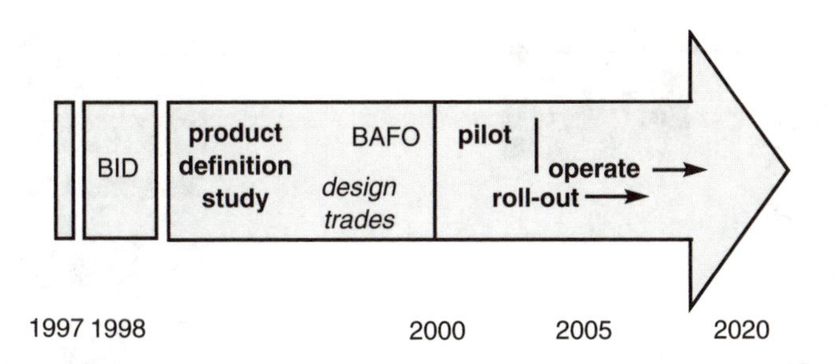

Fig 14.2 The Airwave service timeline.

users, i.e. it will enable the police and public safety organisations to deliver a more effective service to the community.

Key features of the service are geographic rather than demographic radio coverage, provision of a 'private' network designed to meet peak loads and maintain grade of service in times of national emergency, and a highly resilient design.

Applications such as automatic vehicle location (AVL) will be available for better management of resources and safety of users as well as integration with police IT systems such as the Police National Computer and local 'command-and-control' systems.

A key improvement over present systems is seamless communications (Fig 14.3) over organisational and geographic boundaries, enabling high levels of interoperability and therefore a more co-ordinated response to major incidents such as fires, demonstrations, aircraft crashes, etc.

When roll-out is completed in 2005, the Airwave service will utilise approximately 3000 to 3500 radio base-stations providing a very high level of radio coverage for the emergency services and public safety organisations throughout Britain. The base-stations are connected via a network of landlines to a number of switches, and from there to customer control rooms, using BT leased lines.

Public safety users will operate in the 380-400 MHz bands (in the UK this may be limited to 380-383 MHz paired with 390-393 MHz). Throughout Europe, TETRA services may be harmonised in the 870-888 MHz paired with 915-933 MHz bands (when the UK TACS service is closed). Commercial public access mobile radio (PAMR) use is in the 410-430 MHz bands. Private TETRA users (e.g. for industrial site communications) may use the 450-470 MHz band some time after 2006.

TETRA provides voice services and both circuit- and packet-mode data services. Data service bandwidth can be supplied 'on demand' up to 28.8 kbit/s in circuit-switched mode or up to 16 kbit/s in packet mode. The packet mode is a similar service to GSM GPRS, although the maximum data rate capability is lower. New standards work is being undertaken to enhance these rates and is discussed later in this chapter. The Airwave service provides packet-mode operation for its efficiency in working with various applications.

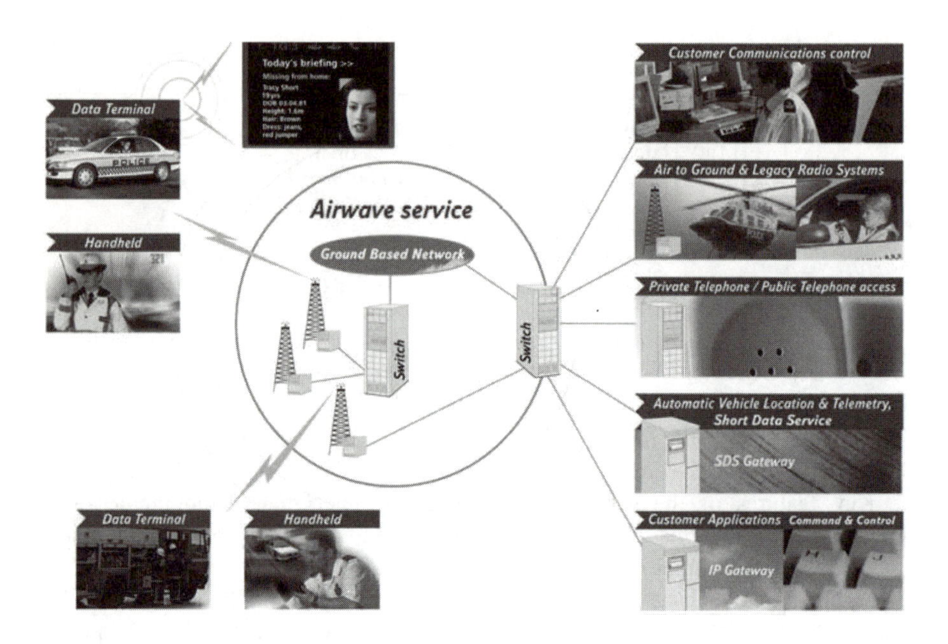

Fig 14.3 The Airwave service will provide seamless communications.

Speech is digitally coded at just 4.8 kbit/s to maximise use of radio spectrum. The ACELP coding scheme is designed to minimise the effect of background noise on speech quality.

In addition the speech signals are encrypted over the air interface as standard, preventing eavesdropping (a problem with analogue systems). Speech may also be encrypted end-to-end if required.

Group calls and broadcast calls are supported with fast call set-up times, typically around 300 ms. Point-to-point calls can be made privately between individuals and to public networks. Priority calls can be set up in emergency situations. Airwave supports gateways into other systems such as PSTN/PTN interconnect, voicemail, and interoperability with older radio systems.

Data calls may be made between mobile terminals or to and from central systems and databases, e.g. automatic resource despatch, automatic vehicle and person location, image and video transfer, database queries.

Where no radio infrastructure is present, terminals may operate in direct mode operation (DMO), i.e. mobile to mobile. Repeaters and gateways are defined to extend the DMO range and get back into the trunked infrastructure respectively (coverage extension) (Fig 14.4).

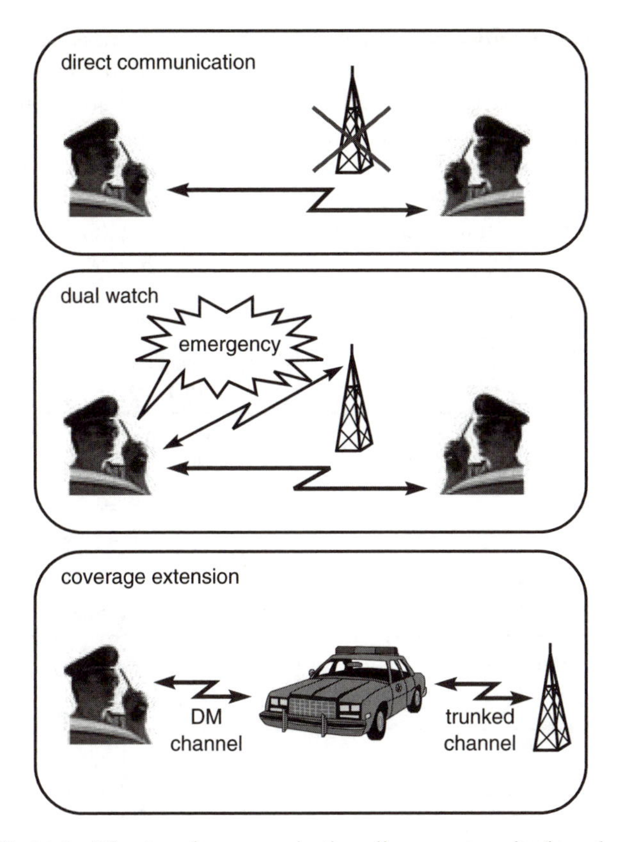

Fig 14.4 Direct mode communication allows communication when no radio infrastructure is present.

BT is responsible for procuring, installing, maintaining and managing the entire Airwave network via a number of network and service centres. The service is offered to users as guaranteed 'core' or 'foundation' plus a menu offering other services for additional coverage, capacity, gateways, etc. This service approach is quite different from that of the public cellular service in the UK.

14.3 Process and Knowledge Working

A useful view of mobile business user requirements can be gained through an analysis of the way people work. The ability of various technologies to meet these requirements can then be examined.

The mobile community can be divided into three classes of user [2, 3]. These are 'knowledge', 'process' and 'residential' users. Residential customers use mobile communications in a casual way, such as in making social arrangements, gaining travel information and just chatting! However, the mobile business community is more purposeful and can be said to have just knowledge and process workers. Knowledge workers use mobile communications to provide information on which they can choose to act, and process workers are driven by a high degree of central control (Fig 14.5).

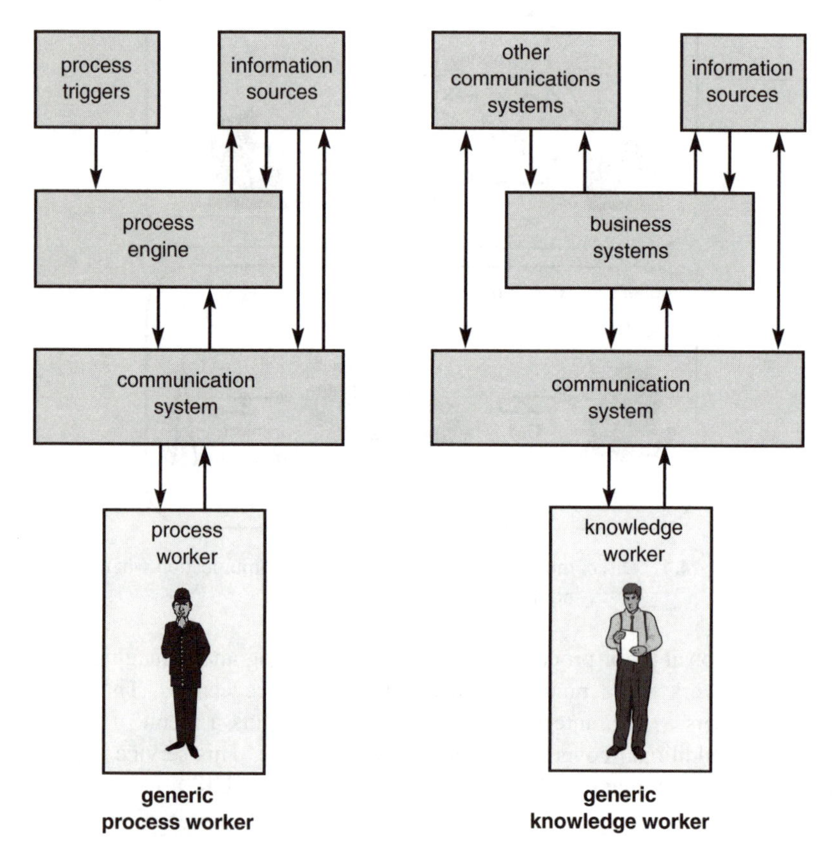

Fig 14.5 Knowledge and process workers.

The PMR TETRA functionality that was listed in section 14.2 is ideally matched to 'process' working business requirements, to which users work within defined business processes. The performance required by public safety users is also a key factor in the choice of TETRA. However, it is recognised that during the development of 'TETRA Release 1' an Internet 'explosion' has occurred, affecting all market-places, setting a new requirement for high-speed data in a 3G mobile telecommunications environment.

The capability of carrying high-speed data and the provision of other key enhancements may allow the PMR market-place to also address the 'knowledge' working business market, where the use of applications is much more *ad hoc*. This high-speed data capability is being addressed through TETRA enhancements, but it is also recognised that possible interworking or integration with other 3G cellular services may also provide benefit. These extended terms of reference for TETRA have been agreed for Release 2 [4].

As an important aspect of the Airwave service is managed security, any inter-working with other 3G services will also have to be made secure.

Although process workers may currently use intranet access, this will be as part of a specific business process, for example, updating plans for the location of pipes and plant for a water company.

14.4 How the Technology may be Evolved to Give Enhanced Capability

The Airwave service has been carefully designed to meet the very exacting requirements of the UK police and other major public safety users, such as fire and ambulance services. These requirements were largely based on centralised 'command and control' working practices that are at present essential to the management of resources using a despatcher, usually operating from a control room. As discussed earlier, this working practice can usefully be described as 'process' working. Process working will continue for many users, with the despatcher process becoming more automated in future using a combination of location services and status messaging with voice as the 'killer' application.

The present Airwave service will meet the public safety users' requirements well, but has additional capability, such as direct voice access from 'mobiles' to the police national computer and force databases, and potential access to other services that will allow changes to more efficient working practices in the future.

While present TETRA technology has high capability, and only PMR technology will provide the functionality and performance required by public safety workers, the benefits of the 'Internet explosion' and the advances in other mobile technologies such as GSM, HSCSD, GPRS and UMTS cannot be ignored. Also, the substitution of traditional circuit-switched core networks with IP networks offer the potential for both cheaper voice transmission and higher instantaneous bandwidth data delivery.

'Process' working practice makes very efficient use of bandwidth, e.g. a status message of maybe a couple of bytes in length can be used to say 'on duty' or 'arrived at incident'. This is appropriate to say a 'beat' officer, but may have less value for a detective, where the information sent and received may not be in a simple format and therefore be less bandwidth efficient.

An example of non-standard information would be the use of electronic witness and interview statements. It is of course vital that security and continuity of evidence (i.e. whether the evidence could have been tampered with during transmission to various parties) are maintained (Fig 14.6). The secure techniques being developed for electronic commerce could well be appropriate to this application, but are not optimised for low-bandwidth transmission. We are moving into an age of digital signatures, where the timely availability of crime/incident reports will lead to more effective policing, e.g. there would be a large benefit in linking 4stolen car report to a robbery report almost instantaneously.

Fig 14.6 The transmission of non-standard information, such as witness statements and pictures, becomes easier with greater bandwidth.

Obviously speech is available for making non-routine enquiries and passing information to and from field workers, but having the bandwidth and connectivity to directly seek and share electronically held information would be of benefit. At a local level, this could be direct access to networked office-based applications such as electronic calendars and briefings (Fig 14.7). This type of access could produce a revolution in the current working practice, from 'process' to 'knowledge' based working, with workers being deployed where their work takes them.

A cardinal requirement of PMR communication is that people can be organised in functional and geographic work groups; this mapping is ideally suited to the sharing of common information. Taking the above argument to the extreme could mean that centralised accommodation such as at police stations would no longer be required (the custody suite excepted!). It may also enable a return to more community-based policing, i.e. working from home. However, a more realistic

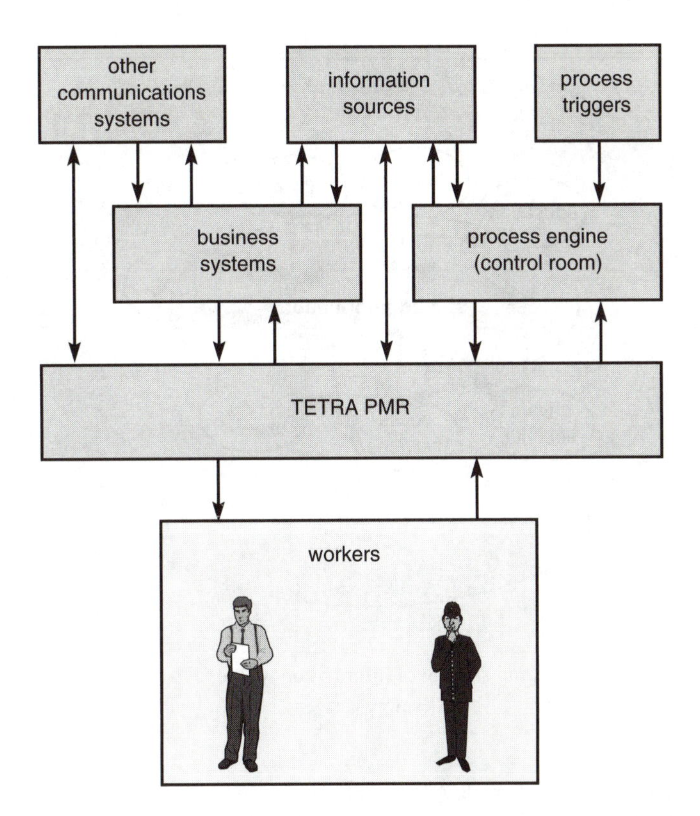

Fig 14.7 In the future, public safety workers may be connected to office-based applications, as well as the control room.

advantage may be the delegation of certain command-and-control functions from the centralised despatcher to operational supervisors in the field. Being a much more efficient and data-rich communications medium compared to existing systems, the new service may enable relevant operational information to be presented directly to the 'front line'.

As said before, knowledge working requires higher data rates, simply because the information being passed will often be in non-standard formats and amount; however, it may also require efficient and highly secure connection to other networks and services, i.e. interworking, allowing information to be 'knitted' together. The need for interworking may be high, e.g. in the need to share information between the Crown Prosecution Service (CPS), Child Protection Agency (CPA), voluntary organisations, hospitals, schools, private security firms and safety experts. Advanced communication interworking will lead to a more effective public safety service (Fig 14.8).

The TETRA fraternity has recognised that the standard will require continued development to meet changing market needs. A key part of this is higher data rates,

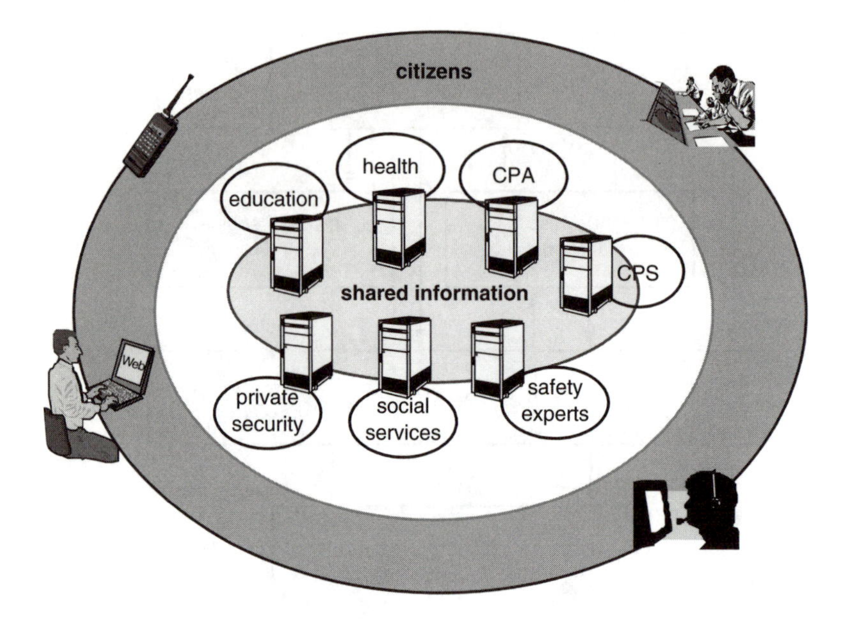

Fig 14.8 Interworking and sharing of information will lead to a more effective public safety service.

but interworking and other developments are also important. The ETSI board recently approved new terms of reference to develop a set of ETSI deliverables that support the development of TETRA in the global competitive 3G market-place [4]. The following are some of the key requirements that have been identified closely aligning with the future Airwave service vision discussed above:

- high-speed packet data in support of multimedia and other high-speed applications;
- additional speech codecs to allow interconnection to other 3G networks without transcoding and to provide enhanced voice quality;
- further enhancements to the air interface to provide increased benefits in terms of spectrum efficiency, subscriber capacity, system performance, quality of service, size and cost of terminals, etc;
- production and/or adoption of standards to provide improved interworking and roaming between TETRA and public mobile networks, such as GSM, GPRS and UMTS;
- evolution of the TETRA SIM, with the aim of convergence with the universal SIM (USIM), to meet the needs for TETRA-specific services while gaining the

benefits of interworking and roaming with public mobile networks, such as GSM, GPRS and UMTS.

BT Airwave is carefully considering how its business will develop in a 3G world, even as the initial pilot service in Lancashire is being 'put through the hoops' by the police and fire services. BT Airwave is part of the newly formed BT Wireless organisation that includes all BT's mobile interests. There is much in common between the various internal businesses and there is much to be gained by sharing work on new architectures, such as integrated core IP networks and the potential for interworking between networks.

This will open up a raft of new service offerings and benefits to 'mobile business communication' users.

14.5 Conclusions

Although PMR has a lower profile than its public cellular counterpart, it is used widely for mobile business communications, by an increasing number of users requiring specialised mobile radio solutions and is the technology of choice for process workers.

The TETRA PMR radio standard offers unique PMR functionality and performance, and is the radio technology adopted for supply of a public safety service to the UK police and other emergency organisations.

Mobile business communications operations can usefully be separated into 'process' and 'knowledge' working categories. A major difference between these two modes of working is the bandwidth (or high-speed data) required for flexible 'knowledge' working. This is recognised by the TETRA fraternity, and work has been agreed to address this within the TETRA Release 2 set of standards.

Another aspect being addressed within TETRA Release 2 is interworking with other 3G services. While this may provide additional benefit, it must not compromise the secure service provided by Airwave.

It is envisaged that both present and future technologies will allow major changes in the way public safety users are deployed, allowing better use of resources. It is also envisaged that the sharing of information between services could realise quicker and more effective public safety services.

BT Airwave is committed to developing its service, which will benefit by using the 'spin-offs' from a 3G world. However, the public 3G world will also have much to learn from how 'mission-critical' mobile business communications are managed by BT Airwave.

Management of procurement and service risks through the increasing use of PFI/PPP, in terms of public safety networks, means investment can be directed into providing a service against a guaranteed revenue stream.

References

1 BT Airwave Web site — http://www.bt.com/airwave

2 Woolf, P.: '*Data over mobile*', Intellact report (March 1999).

3 Tattersall, P. R., and Cushing, A.: '*Mobile Business Communications: User Requirements and Network Solutions*', AES2000 (2000).

4 ETSI Board 28, Sophia Antipolis (September 2000) — http://www.etsi.org/etsiboard

ACRONYMNS

2G	second generation (mobile technology)
3G	third generation (mobile technology)
3GPP	Third Generation Partnership Project
AAA	authorisation, authentication and accounting
AAL	ATM adaptation layer
AAR	allocating access router
ACELP	adaptive code excited linear prediction
ADSL	asymmetric digital subscriber line
AMLCD	active matrix liquid crystal display
AMR	adaptive multi-rate
ANSI	American National Standards Institute
API	application programming interface
AR	access router
ARIB	Association of Radio Industries and Businesses
ARPU	average revenue per user
AS	application server
ASP	application service provision
ATM	asynchronous transfer mode
AVL	automatic vehicle location
BAFO	best and final offer
BBM	break before make
BCSM	basic call state machine
BS	base-station
BSC	base-station controller
BSS	base-station subsystem

BU	binding update
BUAck	binding update acknowledgement
CAMEL	customised applications for mobile network enhanced logic
CAP	CAMEL application part
CC/PP	composite capabilities/preference profiles
CCoA	collocated care of address
CDMA	code division multiple access
CGI	common gateway interface
cHTML	compact HTML
CIDR	classless Internet domain routeing
CLDC	connected, limited device configuration
CN	correspondent node
CoA	care of address
Codec	coder/decoder
COO	cell of origin
COPS	common open policy service
CoR/FA	care of router/foreign agent
CORBA	Common Object Request Broker Architecture
CPA	Child Protection Agency
CPL	call processing language
CPS	Crown Prosecution Service
CRCoA	collocated roaming care of address
CS	circuit-switched
CSCF	call state control function
CSD	circuit-switched data
CWTS	Chinese Wireless Telecommunication Standard
DAG	directed acyclic graph
DAMPS	digital advanced mobile phone system
DECT	digital enhanced cordless telecommunications
DHCP	dynamic host configuration protocol
DID	destination ID
DMO	direct mode operation

DNS	domain name system
DSP	digital signal processing
EDGE	enhanced data rates for GSM evolution
EIR	equipment identification register
EMA	edge mobility architecture
E-OTD	enhanced observed time difference
ETSI	European Telecommunications Standards Institute
FA	foreign agent
FDD	frequency division duplex
GGSN	gateway GPRS support node
GoS	grade of service
GPRS	general packet radio service
GPS	global positioning system
GSM	Global System for Mobile Communications
GSM-AMR	GSM adaptive multi-rate
GSM-EFR	GSM enhanced full rate
GSM-FR	GSM full rate
GUIS	generic user interaction SCF
HA	home agent
HAck	hand-off acknowledgement
HALO	high altitude low orbit
HAVI	home audio visual
HD	hand-off denial
HH	hand-off hint
H-HR	host hand-off request
HI	hand-off initiation
HLR	home location register
HR	hand-off request
HSCSD	high-speed circuit-switched data
HSS	home subscriber server
H-TIN	host tunnel initiation
HTML	hypertext markup language

I-CSCF	interrogating CSCF
IAB	Internet Architecture Board
IETF	Internet Engineering Task Force
IM	IP multimedia
IMSI	international mobile subscriber identity
IP	Internet protocol
IPCC	Inter-governmental Panel on Climate Change
IPVPN	IP virtual private networks
IRC	Internet relay chat
IS-IS	intermediate server to intermediate server
ISUP	integrated services user part
ITU	International Telecommunication Union
IVR	interactive voice response
J2ME	Java 2 micro-edition
JAIN	Java integrated network
JES	Java enhanced SIP
LA	location area
LAN	local area network
LCD	liquid crystal display
LDAP	lightweight directory access protocol
LEAP	lightweight efficient application protocol
LEO	low earth orbit
LFS	location fixing system
LMU	location measurement unit
LPG	liquid propane gas
M3A	mobile multimedia access
MAC	media access control
MANET	mobile *ad hoc* network
MAP	mobile application part
MBB	make before break
MER	mobile enhanced routeing
MExE	mobile execution environment

MGCF	media gateway control function
MGW	media gateway
MH	mobile host
MID	mobile information device
MIP	mobile IP
M-ISP	mobile Internet service provider
MMS	multimedia messaging service
MPLS	multi-protocol label switching
MS	mobile station
MSC	mobile switching centre
MSISDN	mobile station ISDN number
MVNO	mobile virtual network operator
MWIF	Mobile Wireless Internet Forum
NAI	network address identifier
NAR	new access router
NiCd	nickel cadmium
NID	node ID
Ni-MH	nickel metal hydride
NM	network management
NMT	network management terminal
NO	network operator
NRPB	National Radiological Protection Board
OAR	old access router
OFDM	orthogonal frequency division multiplexing
OLO	other licensed operator
OSA	open systems access
OSPF	open shortest path first
O-TDOA	observed time difference of arrival
P-CSCF	proxy CSCF
PAMR	public access mobile radio
PBX	private branch exchange
PCMCIA	Personal Computer Memory Card International Association

PCS	personal communication system
PDA	personal digital assistant
PDC	Personal Digital Cellular (standard) (Japan)
PEST	political, economic, social, and technological
PFI/PPP	private finance initiative/public private partnership
PHS	personal HandyPhone system
PITO	Police Information Technology Organisation
PMLCD	passive matrix liquid crystal display
PMR	professional (private) mobile radio
PS	packet-switched
PSTN	public switched telephone network
QoS	quality of service
RANAP	radio access network application protocol (3GPP)
RFQ	request for quotation
RID	router ID
RMI	remote method invocation
RNC	radio network controller
RSVP	resource reservation protocol
RTCP	RTP control protocol
RTOS	real-time operating system
RTP	real-time transport protocol
SCF	service capability feature
SCS	service capability server
S-CSCF	serving CSCF
SCTP	stream control transmission protocol
SDR	software defined radio
SGSN	serving GPRS support node
SIM	subscriber identity module
SIP	session initiation protocol
SMS	short message service
SP	service provider
SPD	serving profile database

SRNS	serving radio network subsystem (GPRS)
TACS	total access communications system
TCP	transmission control protocol
TD-CDMA	time division — code division multiple access
TDD	time division duplex
TDMA	time division multiple access
TD-SCDMA	time division — synchronous code division multiple access
TETRA	terrestrial trunked radio
TFO	tandem-free operation
TFT LCD	thin film transistor liquid crystal display
TIA	Telecommunications Industry Association
T-IMSI	temporary IMSI
TIN	tunnel initiation
TIPHON	Telecommunications Internet Protocol Harmonisation Over Networks (ETSI)
TOA	time of arrival
TORA	temporally ordered routing algorithm
TrFO	transcoder-free operation
TTA	Telecommunications Technology Association (Korea)
TTC	Telecommunication Technical Committee (Japan)
UD	unicast-directed update
UDP	user datagram protocol
UDU	unicast-directed update
UDUA	UDU acknowledgement
UMTS	universal mobile telecommunications system
USIM	universal SIM
USSD	unstructured supplementary services data
UTRA	UMTS terrestrial radio access
UTRAN	UMTS terrestrial radio access network
VASP	value added service provider
VHE	virtual home environment
VLR	visitor location register

VMSC	visited MSC
VoIP	voice over IP
W3C	World Wide Web Consortium
WAE	wireless application environment
WAP	wireless application protocol
WARC	World Administrative Radio Conference
W-CDMA	wideband — code division multiple access (3GPP)
WDP	wireless datagram protocol
WHO	World Health Organisation
WIM	wireless identity module
WISP	worldwide Internet service provision
WRED	weighted random early drop
WSP	wireless session layer
WT	wireless telegraphy
WTA	wireless telephony architecture
WTLS	wireless transport layer security
WTP	wireless transport layer
WWW	World Wide Web
XML	extended markup language

INDEX